SKILLS Coach
America's Best for Student Success

Write Math!

How to Construct Responses to Open-Ended Math Questions

LEVEL **D**

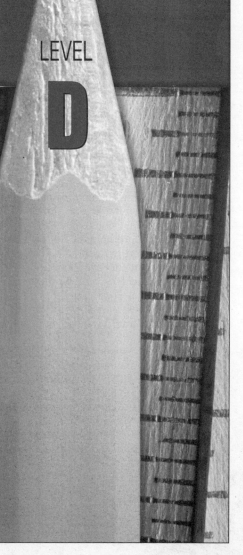

Write Math!
How to Construct Responses to Open-Ended Math Questions, Level D
99NA
ISBN-10: 1-58620-910-8
ISBN-13: 978-1-58620-910-0

EVP, Publisher: Linda Sanford
VP, Editorial Director: Marie Spano
VP of Production: Dina Goren
VP, Creative Director: Rosanne Guararra
Art Director: Farzana Razak

Senior Development Editor: Elizabeth Jaffe
Contributing Author: Keith Grober
Designer: Electric Pictures
Layout artist: Hyun Kounne
Illustrator: Terri Sirell
Cover Design: Farzana Razak
Cover Photo: Myron/The Image Bank/Getty Images

Triumph Learning® 136 Madison, 7th Floor, New York, NY 10016
© 2006 Triumph Learning, LLC
A Haights Cross Communications, Inc. company

Printed in the United States of America.

10 9 8 7 6

Table of Contents

Dear Student,

Welcome to the smart way to write answers to open-ended math questions. You will learn what open-ended math questions are, how to solve them, and how to score them. You will do this by reviewing modeled problems, practicing with guided questions, and answering independent problems. You will work together with your teacher, with your classmates, and with your caregiver at home.

Let's learn to **write math** the smart way!

Have fun!

1. What Is an Open-Ended Math Question?

It is a math problem with a correct answer that you can get to in different ways.

Each way is great as long as:

- **it gets you to the right answer**

- **you show how it got you to the answer**

- **you explain why you chose to answer the question this way**

5-step plan:

1. Read and Think • **2. Select a Strategy** • **3. Solve** • **4. Write/Explain** • **5. Reflect**

This plan will help you answer open-ended questions.

1. **Read and Think**
2. **Select a Strategy**
3. **Solve**
4. **Write/Explain**
5. **Reflect**

Let's take a look at this 5-step plan:

1. Read and Think

Read the **problem** carefully.

What **question** are you asked to find?

● In your own words, tell what this problem is about.

What are the **keywords**?

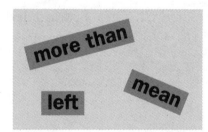

more than

mean

left

What **facts** are you given?

● Decide what facts are needed, and which ones are extra.

2. Select a Strategy

- **How am I going to solve this problem?**

- **What strategy should I use?**

There are lots of strategies to help you solve open-ended math questions. Some are listed below and others you may come up with yourself. In parentheses, you will find where to locate examples of the strategies in use.

Draw a Picture or Graph . . .

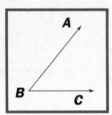

when you need to see the information given in a problem.

(see Chapter 3, Modeled Problems, *p. 19.*)

Make a Model or Act It Out . . .

when you need to watch how the solution is found.

(see Chapter 2, *p. 15.*)

Make an Organized List or Table . . .

1. _____
2. _____
3. _____

when there is a lot of information scattered throughout a problem. A list or table can help you organize your thinking.

(see Chapter 5, Modeled Problem, *p. 31*; or Chapter 5, Guided Problem #1, *p. 33.*)

Look for a Pattern . . .

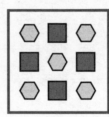

when you need to predict what comes next or find a rule. Making a list or table can often help you find a pattern.

(see Chapter 5, Modeled Problem, *p. 31.*)

Guess and Test . . .

when it is difficult to work out the answer to a problem. Make a guess. Then test it. If your guess is incorrect, use that guess to make a better guess.

(see Chapter 5, Guided Problem #3 *p. 40*; or Chapter 6, Modeled Problem, *p. 47.*)

Logical Thinking . . .

when you need to figure out how the information you have fits together.

(see Chapter 5, Guided Problem #1, *p. 33*, or #2, *p. 36*.)

Work Backward . . .

when you know the end result or total and need to find a missing part.

(see Chapter 6, Guided Problem #1, *p. 41*.)

Write a Number Sentence or Equation . . .

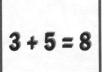

3 + 5 = 8

when you need to find a missing amount or show your work.

(see Chapter 6, Guided Problem #2, *p. 53*.)

Divide and Conquer . . .

$3 \times 4 \div 2 = $?

$3 \times 4 = 12$

$12 \div 2 = $ 6

when you must solve more than one problem to find the answer to the main question. Break down the main question into steps, and solve steps one at a time. Name the strategy you may use for each step.

(see Chapter 6, Guided Problem #1, *p. 50*.)

Make it Simpler . . .

$20 + 40 + 50 = $?

$2 + 4 + 5 = 11$

Then add a zero: 110

when you must solve a complex problem with large numbers or many items.

Reduce the large numbers to small numbers, or reduce the number of items given.

(see Chapter 9, Guided Problem #3, *p. 107*.)

3. Solve

After you pick your strategy, use it to solve the problem.

Use your knowledge of mathematics here.

- **Be very careful—you must get the correct answer. Check your arithmetic!**
- **Label your work. Use units or a sentence that explains the answer.**

4. Write/Explain

Write out an **explanation** of how you solved the problem.

Explain the strategy you chose and why you chose it.

Write your thoughts or why you solved it that way.

Don't leave out any steps.

- **If you came up with a strategy of your own that is not on the list on *pages 8–9*, be sure to explain what the strategy is and why you chose it!**
- **Your writing must be clear. This is very important when you are taking a test. Remember, the person who reads your work must be able to figure out what you did. You can lose points if your writing is not very clear.**

5. Reflect

First, Review Your Work

Use the list below to help you review.

Read It and Think

- ☐ **Did I read the problem at least twice? Do I understand it?**
- ☐ **Did I write down the question being asked?**
- ☐ **Did I write down the keywords in the problem?**
- ☐ **Did I write down the facts that are given?**
- ☐ **Did I write down the strategy that I used?**
- ☐ **Did I solve the problem?**
- ☐ **Is my arithmetic correct?**
- ☐ **Did I explain how I solved the problem?**
- ☐ **Did I explain why I chose the strategy and how I used it?**
- ☐ **Did I include all the steps I took to solve it?**
- ☐ **Is my writing clear?**
- ☐ **Did I label my work?**
- ☐ **Does my answer make sense?**
- ☐ **Did I answer the exact question being asked?**

Then, Improve What You Wrote

How can you improve your writing?

● **Try to rewrite your answer to make it clearer, more accurate, and more complete.**

These steps might look like a lot of work. Once you start to use them, they will become very familiar and not seem so hard. You will find that these five steps work with any open-ended math question. This book will help you practice using them.

Working with this book will help you do better on math tests with open-ended questions. You will even learn how to check your answer using the **rubric** on *page 13*.

The Glossary

A Glossary is like a dictionary. It tells the meaning of words. You should learn to use the Glossary found on *pages 139–152*. It contains mathematical words you should know, including **keywords** found in the math problems throughout this book.

Sometimes a word can mean one thing in everyday life, but something else in mathematics. For example, in everyday life, the word **straight** can mean *immediately*, as in "After you wash up, go **straight** to bed." But in math, **straight** can describe a *kind of line that is not bent or curved*, as in, "The sides of a square are four **straight** lines."

If you are not sure what a word means mathematically in this book, look it up in the Glossary.

2. What Is a Rubric?

I BET A RRRRRRRUBRIC IS AS FUN AS IT IS TO SAY!

A rubric is a grading system used to score open-ended math questions. The person who scores the answers on your test uses a rubric. A rubric can also be used as a guide in answering open-ended math questions. It lists things that should be found in your answer.

Throughout this book, you will use the rubric in two ways:

1. to **guide** you in answering an open-ended math question. It will remind you to write a correct, clear, complete, and thoughtful answer.

2. to **score** yourself as you double-check that your answer is complete.

Here is a typical rubric. It is used to score your work from **0** to **4**. **4** is a perfect answer.

4
- You showed you knew what the problem asked.
- You showed you knew what facts were given, including keywords.
- You chose a good strategy and used it correctly.
- Your arithmetic or operations were done correctly.
- You got a correct and complete answer and labeled it.
- You wrote a good, clear explanation of why you chose a strategy and how you used it.
- You put in all the steps you used to get to your answer.
- You explained your thinking clearly.

3
- You showed you knew what the problem asked.
- You showed you knew what facts were given, including keywords.
- You chose a good strategy but may not have used it correctly,
  OR
  you may have made an arithmetic error in your work.
- You wrote an explanation of why you chose a strategy and how you used it.
- You might not have used all of the steps to get your answer.
- Your explanation was mostly clear but might not have been entirely complete.

2
- You showed you knew what the problem asked.
- You showed you knew what facts were given, including keywords.
- You chose a good strategy but may not have used it correctly,
  OR
  you may have made an arithmetic error in your work.
- Your answer may not be correct.
- Your explanation may not be complete.
- Your explanation may not be clearly written.

1
- You did not understand what the problem asked, OR you did not know what facts were given.
- You did not select a good strategy or did not apply your strategy correctly.
- You made an arithmetic error in your work.
- Your explanation was not complete or you did not write an explanation.
- Your explanation was not clearly written.

0
- You showed no work at all,
  OR
  the work you showed had nothing to do with the problem.

- You will never get a score of **0** if you start to solve the problem.

- You should always write down what you were asked to find out and what facts were given. This shows that you understood some of the problem, and attempted to solve it.

- Getting used to answering a question using a rubric may seem like a lot of work, but once you start to use it, you will see it as being very helpful. You should practice using a rubric at school and at home.

Let's do an open-ended math question. See how the **5-step plan** fits the rubric to help **YOU** get a **4** on your next answer!

Remember the 5-step plan:

1. Read and Think • **2. Select a Strategy**

3. Solve • **4. Write/Explain** • **5. Reflect**

1. Read and Think

Read the **problem** carefully.

Modeled Problem

Keira and Mick are collecting CDs. Mick collected 6 cases with 4 CDs in **each** case. Keira collected 4 cases. She had 6 CDs in each case. Who has **more** CDs?

Keywords: each, more

What **question** are you asked?

● The question tells you what it is you want to find out. You must answer the question in order to solve the problem correctly.

● Here, the question is, "Who has more CDs?"

What are the **keywords**?

● **each** to consider individually

● **more** greater in number or amount

Check the Glossary on *p. 139*

What **facts** are you given?

● Every problem has facts, data, or information. Facts help you answer the question.

In this problem, the **facts** are:

● Keira collected 4 cases with 6 CDs in each case.
● Mick collected 6 cases with 4 CDs in each case.

2. Select a Strategy

In order to solve a problem, you need to use a **strategy**.

There are many strategies you can use. *Chapter 1, pages 8–9* shows some strategies you might use. You may also choose to use one of your own.

Let's look at how two students chose different strategies to solve this problem.

First Solution	Second Solution

First, we will show what Bobby did.

This involves using a strategy, **Writing a Number Sentence**.

3. Solve

First solution—**Write a Number Sentence**:

Keira's CDs
4 cases with 6 CDs in each case
$4 \times 6 = 24$

Mick's CDs
6 cases with 4 CDs in each case
$6 \times 4 = 24$

Keira and Mick have the same number of CDs.

Second, we will review what Caroline did using a strategy called **Acting It Out**. She used chips for CDs and cups for the cases.

3. Solve

Second solution—**Act It Out**:

Mick has more CDs.

Turn the page to see the end of this problem.

Tip

• Many problems can be solved using different strategies. As long as your choice leads to a correct answer and a correct explanation, it is a good choice!

First Solution	Second Solution

First Solution

4. Write/Explain

You must give a written explanation of how you solved the problem and what you were thinking.

Clearly explain what you did and why you did it.

Do not leave out any steps.

> I multiplied 4 x 6 to find the number of CDs Keira had. I found that Keira had 24 CDs in all. I multiplied 6 x 4 to find that Mick also had 24 CDs in all.
>
> They each had 24 CDs. They had the same number.

5. Reflect

Bobby reviewed his work by checking it against the rubric. He answered the question being asked, chose a good strategy, and used it correctly. His arithmetic was correct. He explained why he chose the strategy and how he solved the problem. Finally, he labeled his work.

Score:

Bobby gets a score of **4**.

Second Solution

4. Write/Explain

> Since Keira collected 4 cases with 6 CDs in each, I took 4 cups and put 6 chips in each cup.
>
> Then I added 6 + 6 + 6 + 6 and got 24.
>
> Since Mick collected 6 cases with 4 CDs in each, I took 6 cups and put 4 chips in each cup.
>
> Then I added 4 + 4 + 4 + 4 + 4 + 4 and got 26.
>
> This shows that Mick has more CDs than Keira.

5. Reflect

Caroline reviewed her work by checking it against the rubric. She answered the question being asked. Caroline chose a good strategy and used it correctly, she explained why she chose the strategy, and how she solved the problem. Caroline labeled her work. However, Caroline made a mathematical error. Her addition total for Mick's CDs is wrong.

Score:

Caroline gets a score of **3**.

She could get a **4** by adding correctly to get that Mick had 24 CDs.

Using the Rubric

Whenever you solve an open-ended math question, you can use the rubric on *page 13* as a guide. Use the list in the **4** box as a checklist. This will remind you what to include in your answer for the highest score possible.

Reviewing Your and a Partner's Work

After you finish solving the question, **self-assess**. This means that you should use the whole rubric to review your work and score it. How well did you do? If you need to raise your score, take the time to do so.

You may also **peer-assess**. Swap your work with a partner. Use the rubric to score each other's solution. Now talk about the different answers and the scores that were given. What might seem clear to you may not be clear to your friend. Partners can help each other learn what should be improved. You can discuss different ways to solve the same problem. The more you talk about your mathematics, the more you will understand how to improve your work.

2-Step Decision-Making Process

Some students find it easier before scoring with a rubric, to first use the **2-Step Decision-Making Process** as seen in the next column. It helps decide if you or your partner's answer is a **3 or 4** or a **1 or 2**.

2-Step Decision-Making Process

Before you use a rubric, use a **2-Step Decision-Making Process**. This will give you a jump on scoring your work or your partner's work.

Decide if your work is:

- *acceptable* (3 or 4)

  or

- *unacceptable* (1 or 2).

If your work is *acceptable*, decide if it is:

- **full and complete (4)**

  or

- **nearly full and complete, but not perfect (3)**.

If your work is *unacceptable*, decide if it shows:

- **limited or only some understanding (2)**

  or

- **little or no understanding (1)**.

A **(0)** is **no attempt**.

Solve questions using the rubric as a guide. You will see improvement in your answers and ability. In time you may no longer need a rubric to guide you. The information that must be included in your answer will become very familiar. Then, you will only need the rubric to score your answer.

3. How to Answer an Open-Ended Math Question

We know what an **open-ended math problem** is. We know how to solve it and how it will be scored. Now let's take a problem and solve it together. Then we will see how other students answered it. We will use the **rubric** to see how well they did. Then we will talk about how they could **improve** their answers.

Modeled Problem

Mary's front lawn is **square**. Each **side** is 30 feet long. Her father wants to put a **border** of rose bushes around the lawn. The rose bushes will be planted 5 feet **apart**, and there will be a rose bush in each **corner**. How **many** rose bushes will be planted?

Keywords: square, side, border, apart, corner, many

1. Read and Think

What **questions** are we asked?

● **How many rose bushes are needed?**

What are the **keywords**?

● **square, side, border, apart, corner, many.**

What **facts** are we given?

● **The lawn is a square with 30-foot sides.**
● **The rose bushes will be planted 5 feet apart.**
● **There will be one rose bush in each corner.**

What is **going on**?

● **We know that the lawn is a square. This means that all four sides are 30 feet long. The family is going to plant rose bushes around the border of the lawn.**

2. Select a Strategy

Let's **Draw a Picture** to see what the lawn and its border look like. **Draw a Picture** is one of our strategies. The lawn is a square, 30 feet on each side.

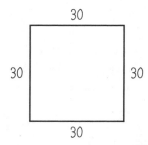

Let's put the corner rose bushes into our drawing.

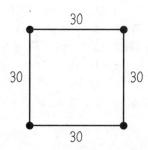

3. How to Answer...

Now let's put the rest of the rose bushes along each side of the border.

They must be 5 feet apart.

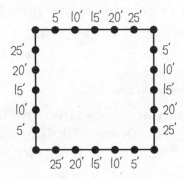

3. Solve

Now that we see what is going on, we can find the answer. Let's count the number of rose bushes. There are 24 rose bushes in all.

4. Write/Explain

We started by **Drawing a Picture** of a square lawn. Since all sides of a square are equal, each of the 4 sides is 30 feet. Next, we put a dot to show the rose bushes in the corners. Then we put dots on each side, making them 5 feet apart. Next we counted the dots. There were 24 dots. So, 24 rose bushes are needed to make the border.

5. Reflect

We reviewed our work. There is one rose bush in each corner. All rose bushes are 5 feet apart. The drawing is clear and correct.

Score:

This solution would earn a perfect **4** on our rubric.

- We showed that we knew what was asked and what information was given, including facts and keywords.
- We chose a good strategy and applied it correctly.
- We did the arithmetic and got the correct answer.
- We wrote a good explanation of how we used our strategy.
- We included all of our steps.
- We labeled our work.
- We clearly explained what we did and why.

Now let's look at some answers that were done by other students.

Tim's Paper

0' 5' 10' 15' 20' 25' 30'

Each side has 7 rose bushes. There are 4 sides to a square, so there are 7 x 4 = 28 rose bushes.

Write/Explain: I drew one side of the square. Then I put a rose bush in each corner. I put the rose bushes 5 feet apart, and counted 7 rose bushes on the side. Since there are 4 sides to the square, and all sides are equal, I multiplied 4 x 7 = 28 to get my answer.

Let's use our rubric to see how well Michael did.

- Did he show that we knew what the problem asked? **Yes.**

- Did he know what the keywords were? **Yes.**

- Did he show that he knew what facts were given? **Yes.**

- Did he name and use the correct strategy? **Yes. He Made a Drawing, but it was only a part of the drawing. He only showed one side of the square.**

- Was his mathematics correct? **No. Michael only drew part of the drawing so did not see that his drawing would give him the wrong answer.**

- Did he label his work? **Yes.**

- Was his answer correct? **No. He got 28 bushes instead of 24.**

- Were all of his steps included? **No. He needed to finish his drawing for this step to be complete.**

- Did he explain why he chose the strategy and how he used it? **Yes.**

- Did he write a good, clear explanation of his work? **Yes.**

Score:

Tim would get a **3** on our rubric.

He did not use the strategy completely. He drew only one side of the square. Tim should have drawn all sides of the square. If he had done that, he would have counted only 24 rose bushes. He did not see that he had counted the bushes in each corner twice.

To get a 4, Tim should have gone back to check his work, drawing all sides of the square. Then he would have seen that he made a mistake in counting.

Ashley's Paper

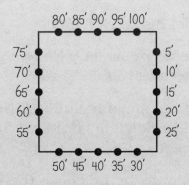

80' 85' 90' 95' 100'

75'   5'
70'   10'
65'   15'
60'   20'
55'   25'

50' 45' 40' 35' 30'

They need 20 rose bushes.

Write/Explain: I Drew a Picture of the square lawn and put in the rose bushes 5 feet apart. Then I counted the rose bushes. They need 20 rose bushes.

- Did she show that she knew what the problem asked? **Yes.**

- Did she know what the keywords were? **No. She missed "corner."**

- Did she show that she knew what facts were given? **No. She missed that there was a rose bush in each corner.**

- Did she name and use the correct strategy? **Yes.**

- Was her mathematics correct? **No.**

- Did she label her work? **Yes.**

- Was her answer correct? **No. She got 20 bushes instead of 24.**

- Were all of her steps included? **Yes.**

- Did she explain why she chose the strategy and how she used it? **Yes.**

- Did she write a good, clear explanation of her work? **No.**

Score

Ashley would receive a **2** on our rubric.

To get a 4, she would have to have redrawn the rose bushes correctly on her diagram, starting with one in each corner.

Tanya's Paper

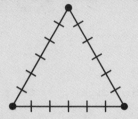

They need 15 rose bushes.

Write/Explain: I Drew a Picture of the lawn and put in the rose bushes 5 feet apart with one in each corner. Then I counted the rose bushes. They need 18 rose bushes.

- Did she show that she knew what the problem asked for? **Yes.**

- Did she know what the keywords were? **No. She missed "square."**

- Did she show that she knew what facts were given? **No. She drew a triangle.**

- Did she name and use the correct strategy, explaining why it was chosen? **Yes.**

- Was her mathematics correct? **No. Drawing a triangle made her math wrong.**

- Did she label her work? **Yes.**

- Was her answer correct? **No. Martha got 18 bushes instead of 24.**

- Were all of her steps included? **Yes.**

- Did she write a good, clear explanation of her work? **No.**

Score

Tanya would receive a **2** on our rubric.

To get a 4, Tanya would have to have drawn a square instead of a triangle, and counted the correct number of rose bushes.

Sean's Paper

4 x 20 = 80

Write/Explain: They need 80 rose bushes.

Score

Sean would receive a **1** on our rubric.

Sean did show the square has 4 sides. But he did not understand the problem and wrote no explanation of his work.

To get a 4, Sean needs to reread the problem to get a better understanding of it and then start from the beginning.

4. How NOT to Get a ZERO!

IT'S OKAY TO HAVE A 0 ON MY JERSEY, BUT NOT ON MY MATH ANSWER!

No one wants to get a **0** on an open-ended math problem. However, you can almost always get some points. The only person who gets a **0** is the person who **leaves the paper blank** or who **writes something that doesn't have anything to do with the problem**. Let's see how we can start by scoring a **1 or 2** on our work, and then bring it up to a **3 or 4**.

And remember:

1. **Read and Think**
2. **Select a Strategy**
3. **Solve**
4. **Write/Explain**
5. **Reflect**

24

How to Get a 1 or 2!

Here is how to get **some credit** on an open-ended math question.

1. **Read** the question. Then **reread** it.

   Ask: "What are the **keywords** to help me solve the problem?"

   Finish the sentence: **"The keyword/s are**

   _____."

You will get credit for listing the keywords.

2. **Understand** the problem. **Repeat the story** of the problem in your own words.

   Ask: "What am I **being asked** to do? What do I **need to find**?"

   Finish the sentence: **"I need to find**

   _____."

You will get credit for listing what was asked.

3. **Find the facts** in the problem.

   Ask: "What does the problem tell me? What do I **know**?"

   Finish the sentence: **"The things I know are**

   _____."

You will get credit for writing the facts.

4. Figure out what **strategy** you will use to help you solve the problem.

   Ask: "What can help me to find what I need to know?

   Finish the sentence:
   "The strategy I will use is

   _____."

You will get credit for listing the strategy you use.

How to Get a 1 or 2
(continued on next page.)

- You will practice these steps as you help solve the modeled problem introduced on the following page.

Tip

- To get points right away, always begin by writing down what you are asked to find and what facts you are given.

Let's do an open-ended math problem together.

First, let's try to get a **1 or 2**.

Modeled Problem

Maria and Leo ordered lunch.
Maria bought a sandwich for
$2.20 and a juice for $0.75.
Leo bought 2 slices of pizza for
$4.25 and a soda for $1.25. How much **more** did
Leo's lunch cost than Maria's?

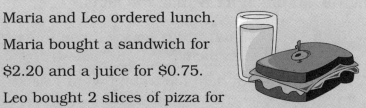

Keyword: more

1. Read and Think

1. Carefully **read** this **question**.
   Reread the question to fully understand it.

2. What is the **keyword** to help you solve
   the problem?

   ● more

   By listing the keyword you can get a
   score of 1.

3. What **question** are you being asked to **find**?

   ● I need to find how much more Leo's lunch
   cost than Maria's lunch.

By writing what you are being asked to find, you
can get a **score of a 1 or 2**.

4. What are the **facts**?

   ● **Maria spent $2.20 on a sandwich.**
   ● **Maria spent $0.75 on a juice.**
   ● **Leo spent $4.25 on pizza.**
   ● **Leo spent $1.25 on a soda.**

By listing the facts, you can get a **score of a 1 or 2**.

2. Select a Strategy

Now you have to pick a strategy and solve the
problem. There are lots of strategies from which to
pick. You may choose one from *pages 8–9* or choose
your own. Your may choose a different strategy
than your classmates. You may solve it differently,
but you both can find the right answer. There is
not only one right way.

1. What **strategy** will you use?

   ● I will use a strategy called **Divide and
   Conquer.** First I will **Make a Table** to organize
   all the facts in the problem. Then I will use
   (continued on page 27)

Write Number Sentences to help me solve the problem and find the answer.

By writing what strategy you chose you can get a **score of 1 or 2**.

Now, let's try to increase our score of the same problem from a **1 or 2** to a **3 or 4**.

How to Get a 3 or 4!

First, let's review. *Remember, you can always get some credit* for listing keywords and the question that is asked. You will also receive points for writing the facts that are given. Finally, credit will be given for listing the strategy you have chosen. By doing this, you will receive a score of at least a **1 or 2**. Now it is time to raise your score to a **3 or 4**. Use all the following information introduced to you to do so.

So here we go: Here is how to take a score of **1 or 2** and make it a **3 or 4**. We will continue using the same modeled problem.

See *pages 8–9* for a list of some strategies you can choose to use.

NOTICE: Photocopying any part of this book is prohibited by law.

3. Solve

Go back to read the problem. How can we find how much more Leo's lunch cost than Maria's? Use the **Divide and Conquer** strategy. *First*, we Make a Table to organize the data.

Name	Leo	Maria
Food	$4.25	$2.20
Drink	$1.25	$0.75
TOTAL	$5.50	$2.95

Then we Write Number Sentences to find out how much each lunch cost.

Leo	Maria
$4.25	$2.20
+ 1.25	+ 0.75
$5.50	$2.95

Leo's lunch cost $5.50 and Maria's cost $2.95. Now we Write another Number Sentence to subtract.

$$\begin{array}{r} \$5.50 \\ -\ 2.95 \\ \hline \$3.55 \end{array}$$

Leo's lunch cost $2.55 more than Maria's lunch.

Tip

- We're going to use a lot of strategies in this book.

- An open-ended math problem can be solved in more than one way. And if your way works out and gives the correct answer, then you are right!

Making the Table helped us see how the facts fit together. **Writing Number Sentences** helped us solve the problem.

Always remember to **label** your work. Use **units** or a sentence that explains what you found.

4. Write/Explain

The person marking your paper does not know what you were thinking. You must explain why you chose the strategy. You should explain how you solved the problem. Your work should be labeled. Don't leave out any steps. Be sure to reread your writing, making sure your work is clear and complete.

Leo and Maria are eating lunch. We had to find how much more Leo's lunch cost than Maria's. We used the Divide and Conquer Strategy. First we Made a Table and labeled it to organize and make sense of all the facts in the problem. These facts included the prices of all the food Leo and Maria ate. Then we Wrote Number Sentences to find how much more Leo's lunch costs than Maria's. First we added the cost of each lunch separately. Leo's lunch cost $5.50 in all. Maria's lunch cost $2.95 in all. Then we subtracted the cost of Maria's lunch from the cost of Leo's lunch. We found out that Leo's lunch cost $3.55 more than Maria's.

5. Reflect

Review Your Work and Improve It!

After you solve the problem, carefully review your work.

- **Did you write what the problem asked you to find?**

- **Did you list all the keywords and facts?**

- **Did you list the strategy you chose to use?**

If you did these things, you will get a score of 1 or 2

- **Did you use the right strategy?**

- **Is your arithmetic right?**

- **Did you label your work?**

- **Did you write out all the steps to solving the problem?**

- **Did you explain why you chose the strategy and how you used it?**

- **Did you explain why you solved the problem the way you did?**

- **Is your writing clear?**

If you did these things, you will raise your score from a 1 or 2 to a 3 or 4.

If you do the things we have suggested, you **CANNOT** get a **0**.

Remember, never leave your paper blank.

Working with Peers

You might want to exchange papers with a friend in class. See if your friend understands what you wrote. That's a good way to see how clearly you explained your work.

Here is a checklist for you to follow. It will make sure you have done your best job. Keep practicing what is on this list. You will improve in solving open-ended mathematics questions.

Read It and Think

- ☐ Did I read the problem at least twice? Do I understand it?
- ☐ Did I write down the question being asked?
- ☐ Did I write down the keywords in the problem?
- ☐ Did I write down the facts that are given?
- ☐ Did I write down the strategy that I used?
- ☐ Did I solve the problem?
- ☐ Is my arithmetic correct?
- ☐ Did I explain how I solved the problem?
- ☐ Did I explain why I chose the strategy and how I used it?
- ☐ Did I include all the steps I took to solve it?
- ☐ Is my writing clear?
- ☐ Did I label my work?
- ☐ Does my answer make sense?
- ☐ Did I answer the exact question being asked?
- ☐ Always check your work!

5. Number and Operations

Numbers are all around us, every day. They make up our **phone numbers**, tell us the **location** of where we live, let us know **how much** something costs, and on and on! We are going to learn more about numbers, explaining how you use them. We will understand using one number alone, how numbers can be used together, how they can be compared, and more.

Here is a problem that you might find on a test. Let's solve it together, to show what a model solution should look like. Then we will score it using our rubric.

Modeled Problem

Beth plans to walk 2 miles **each** day for 2 days. For the next 3 days she will walk 3 miles each day. Then she will walk 4 miles each day for 4 days. She will continue this **pattern** until she is walking 6 miles each day. On what day will she **first** walk 6 miles?

Keywords: each, pattern, first

1. Read and Think

What **question** are we asked?

● We are asked to find out on what day will Beth first walk 6 miles.

What are the **keywords**?

● each, pattern, first

What are the **facts**?

● Beth will walk 2 miles for 2 days.
● Beth will walk 3 miles for 3 days.
● Beth will walk 4 miles for 4 days.
● The pattern will continue.

2. Select a Strategy

There are *two parts* to this problem. For the *first* part, we will use the **Make a Table** strategy to organize the information. For the *second* part, we will use the **Look for a Pattern** strategy to answer the question.

3. Solve

First, we will **Make a Table** to see the information more clearly, helping us make sense of the information. Now we will use the **Look for a Pattern** strategy to find how many days Beth walked 5 miles.

Miles Walked	Number of Days Walked	Days
2	2	1, 2
3	3	3, 4, 5
4	4	6, 7, 8, 9
5	5	10, 11, 12, 13, 14

The first day Beth will walk 6 miles is the 15th day.

4. Write/Explain

We **Made a Table** to show the number of days that Beth walked each distance. We **Used the Pattern** to find that Beth walked 5 miles each day for 5 days. We kept track of the days and found out that the first day she walked 6 miles was the 15th day.

5. Reflect

Let's review our work and answer.

- Did we show that we knew what the problem asked? **Yes. We answered the question that was asked.**

- Did we know what the keywords were? **Yes.**

- Did we show that we knew what facts were given? **Yes.**

- Did we name and use the correct strategy? **Yes.**

- Was our mathematics correct? **Yes. We checked it. It was correct.**

- Did we label our work? **Yes.**

- Was our answer correct? **Yes.**

- Were all of our steps included? **Yes.**

- Did we explain why we chose the strategy and how we used it? **Yes.**

- Did we write a good, clear explanation of our work? **Yes.**

Score

This solution would earn a **4** on our rubric. It was perfect.

Here are some **Guided Open-Ended Math Problems**.

For each problem there are **four parts**. In the **first part**, you will solve the problem with guided help. In the **second part**, you will score and correct a solution with guided help. The **third part** shows one solution that scores a perfect **4**. This solution may or may not differ from your way. The **fourth part** has *answers* to the **first** and **second parts** so you can check your work.

Guided Problem #1

On what dates of the month is the sum of the digits 4?

Keywords:

1. Try It Yourself.

Answer the questions below to get a score of **4**.

What **question** are you being asked?

_____

What are the **keywords**?

_____

_____

What are the **facts** you need to solve the problem?

_____

_____

What **strategy** can you use to solve the problem?

_____

_____

Solve the problem.

_____

_____

_____

_____

_____

> **Hint**
>
> Possible answers include **Make a List** and **Use Logical Thinking**.

Write/Explain what you did to solve the problem.

_____

_____

Reflect. Review and improve your work.

_____

2. Zachary Tries It.

Zachary's Paper

Question: Which numbers from 1 to 30 have digits that have a sum of 4?

Keywords: month, sum, digits

Facts: A month has 30 days.

Strategy: I made a list.

1, 2, 3, **4**, 5, 6, 7, 8, 9, 10, 11, 12, **13**, 14, 15, 16, 17, 18, 19, 20, 21, **22**, 23, 24, 25, 26, 27, 28, 29, 30

There are 3 dates with a sum of 4. They are the 4th, 13th, and 22nd.

Write/Explain: I Made a List of the numbers from 1 to 30 and picked out the three with a sum of 4.

Score the Answer.

According to the rubric, from **1** to **3**, what score would you give Zachary? Explain why you gave that score.

_____

_____

_____

_____

Make it a 4! Rewrite.

_____

_____

_____

_____

_____

_____

_____

Use the rubric on *page 13* to score this work.

3. Sarah Tries It.

Remember, there is often more than one way to solve a problem. Here is how Sarah solved this problem.

Sarah's Paper

Question: Which numbers from 1 to 31 have digits that have a sum of 4?

Keywords: month, sum, digits

Facts: A month can have 31 days.

Strategy: I used Logical Thinking.

Solve: The addends to get a sum of 4 are: 0 + 4, 1 + 3, and 2 + 2.

The numbers from 1 to 31 that work are 4, 13, 22, and 31.

Write/Explain: I used Logical Thinking to find the addends that have a sum of 4. I found each possible way to get a sum of 4. Then I wrote the numbers from 1 to 31 that work.

Score: Sarah's solution would earn a **4** on a test. Sarah identified the question, the keywords, and the facts. She used a good strategy. Then she clearly explained the steps she took to solve the problem. She labeled her answer. It is perfect!

4. Answers to Parts 1 and 2.

Guided Problem #1

On what dates of the month is the sum of the digits 4?

Keywords: month, sum, digits

1. Try It Yourself. (pages 33–34)

Question: Which numbers from 1 to 31 have digits that have a sum of 4?

Facts: A month can have 31 days in it.

Strategy: Make a Table.

Solve:

S	M	T	W	T	F	S
1	2	3	4	5	6	7
8	9	10	11	12	13	14
15	16	17	18	19	20	21
22	23	24	25	26	27	28
29	30	31				

The dates are the 4th, 13th, 22nd, and 31st.

Write/Explain: I **Made a Table** into a calendar of a month. I then added the digits of each number and indicated the ones that have a sum of 4.

2. Zachary Tries It. (page 34)

Score the Answer: I would give Zachary a **3** on his paper. He listed the question, keywords, and facts. He used a good strategy, but he made one mistake. Most months have 31 days. He included only 30 days. He was missing the 31st.

Make it a 4! Rewrite.

Zachary only needed to include 31 in his list of numbers to get a **4**.

Guided Problem #2

How many 3-digit numbers can you make by using 2, 4, and 8, using each digit only once in each number?

Keywords:

1. Try It Yourself.

Answer the questions below to get a score of **4**.

What **question** are you being asked?

_____

_____

What are the **keywords**?

_____

_____

What are the **facts** you need to solve the problem?

_____

_____

What **strategy** can you use to solve the problem?

_____

_____

Possible answers include **Make a List** and **Use Logical Thinking**.

Solve the problem.

_____

_____

_____

_____

Write/Explain what you did to solve the problem.

_____

_____

_____

_____

_____

Reflect. Review and improve your work.

_____

_____

_____

2. Marsha Tries It.

Marsha's Paper

Question: How many 3-digit numbers can I make with 2, 4, and 8?

Keywords: 3-digit numbers, once

Facts: Each number will have only 2, 4, or 8 in it.

Strategy: I Made a List.

Solve: 222, 224, 228, 242, 244, 248, 282, 284, 288
422, 424, 428, 442, 444, 448, 482, 484, 488
822, 824, 828, 842, 844, 848, 882, 884, 888

There are 27 three-digit numbers that can be made from 2, 4, and 8.

Write/Explain: I Made a List and made as many different 3-digit numbers as I could using 2, 4, and 8. Then I counted the number of numbers and there were 27.

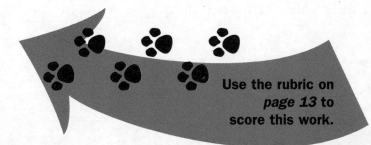

Use the rubric on **page 13** to score this work.

Score the Answer. According to the rubric, from **1** to **3**, what score would you give Marsha? Explain why you gave that score.

_____

_____

_____

Make it a 4! Rewrite.

_____

_____

_____

_____

_____

_____

3. Emily Tries It.

Remember, there is often more than one way to solve a problem. In the next column, see how Emily solved this problem.

Emily's Paper

Question: How many 3-digit numbers can I make using the digits 2, 4, and 8, while using each digit once?

Keywords: 3-digit number, once

Facts: Each number will have 3 digits using the digits 2, 4, and 8.

Strategy: I used Logical Thinking.

Solve: I made a list of all of the numbers that start with 2: 248, 284.

Since there were 2 numbers that start with 2 in the hundreds place, and 3 numbers that could go in the hundreds place, I multiplied 2 x 3 = 6.

There are 6 three-digit numbers.

Write/Explain: I used Logical Thinking. I found the numbers that had 2 in the hundreds place. There were 2 numbers. Since there are 3 different digits, I multiplied 2 x 3 = 6 to find how many 3-digit numbers I could make using 3 digits.

Score: Emily's solution would earn a **4** on the test. She identified the question that was asked, the keywords, and the facts. She picked a good strategy and used it correctly. Then she clearly explained the steps she had used to solve the problem and labeled her answer. It is perfect!

4. Answers to Parts 1 and 2.

Guided Problem #2

How many 3-digit numbers can you make by using 2, 4, and 8, using each digit only once in each number?

Keywords: 3-digit number, digit, once

1. Try It Yourself. (pages 36–37)

Question: How many 3-digit numbers can I make using the digits 2, 4, and 8, while using each digit once?

Facts: Each number will have 3 digits using the digits 2, 4, and 8.

Strategy: Make a List.

Solve: Make a List from least to greatest.

248, 284, 428, 482, 824, 842

There are 6 three-digit numbers that can be made.

Write/Explain: I **Made a List** of the 3-digit numbers using 2, 4, and 8 exactly once. There were 6 numbers.

2. Marsha Tries It. (pages 37–38)

Score the Answer: I would give Marsha a **2** on the rubric. She gave the keywords and her math was correct. However, she did not understand that each digit could only be used once. This led her to answer a different question than the one that was asked.

Make it a 4! Rewrite.

Marsha needs to find only those numbers with 2, 4, and 8 included one time each.

248, 284, 428, 482, 824, 842

Guided Problem #3

Lucy and Marcus filled 24 bags with aluminum cans to recycle. Lucy filled twice as many bags as Marcus. How many bags of cans did each of them fill?

Keywords:

Answer the questions below to get a score of **4**.

What **questions** are you being asked?

_____

_____

What are the **keywords**?

_____

_____

What are the **facts** you need to solve the problem?

_____

_____

What **strategy** can you use to solve the problem?

_____

_____

Solve the problem.

_____

_____

_____

Hint

Possible answers include **Draw a Picture** and **Guess and Test**.

Write/Explain what you did to solve the problem.

_____

_____

Reflect. Review and improve your work.

_____

_____

2. Miguel Tries It.

Miguel's Paper

Marcus filled 12 bags and Lucy filled 6 bags.
Write/Explain: I used Guess and Test. I guessed 12 for Mark and 6 for Lucy. I tested my guess. That worked since 12 is twice as many as 6.

Score the Answer.

According to the rubric, from **1** to **3**, what score would you give Miguel? Explain why you gave that score.

_____

_____

_____

_____

_____

Make it a 4! Rewrite.

_____

_____

_____

_____

_____

_____

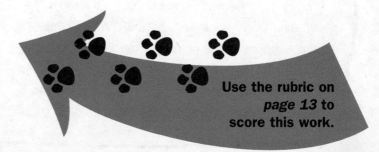

Use the rubric on *page 13* to score this work.

3. Paula Tries It.

Remember, there is often more than one way to solve a problem. Here is how Paula solved this problem.

Paula's Paper

Questions: How many bags did Lucy fill?
How many bags did Marcus fill?

Keywords: twice as many

Facts: They filled 24 bags. Lucy filled twice as many bags as Marcus.

Strategies: I Made a Table and I used Guess and Test

Solve:

Lucy	Marcus	Total
10	5	15
12	6	18
14	7	21
16	8	24

Lucy filled 16 bags and Mark filled 8.

Write/Explain: For the first step, I used the Guess and Test strategy. I Made a Table to keep my guessing work organized. I tried combinations of numbers where one number was double the other. I Guess-and-Tested until I reached 16 for Lucy and 8 for Marcus. I added 16 + 8 = 24. So Lucy had 16 bags and Marcus had 8 bags for a total of 24 bags.

Score: Paula would get a **4** on our rubric. She knew what question was asked and wrote the keywords and the facts. She used a good strategy, and her math was correct. Her answer was correct and she showed all her work. She labeled her answer.

4. Answers to Parts 1 and 2.

Guided Problem #3

Lucy and Marcus filled 24 bags with aluminum cans to recycle. Lucy filled twice as many bags as Marcus. How many bags of cans did each of them fill?

Keywords: twice as many

1. Try It Yourself. (pages 40–41)

Questions: How many bags did Lucy fill? How many bags did Marcus fill?

Facts: They filled 24 bags all together. Lucy filled twice as many as Marcus.

Strategy: Draw a Picture.

Solve: There are 24 bags, and for every 2 that Lucy fills, Marcus fills one.

L L M   L L M   L L M   L L M
L L M   L L M   L L M   L L M

Lucy filled 16 bags and Marcus filled 8 bags.

Write/Explain: I **Drew a Picture**. I drew 24 letters to represent who filled which bag. Since Lucy filled twice as many as Marcus, I drew 2 L's for every 1 M. I then counted the letters and there were 16 L's and 8 M's. Lucy filled 16 bags and Marcus filled 8 bags.

2. Miguel Tries It. (page 41)

Score the Answer: I would give Miguel a **1** on his paper. He did not give the question that was asked, the keywords, or the facts. Miguel did try to answer the question. He used a good strategy and labeled his answer, but did not get the correct answer.

Make it a 4! Rewrite.

Questions: How many bags did Lucy fill? How many bags did Marcus fill?

Keywords: twice as many

Facts: They filled 24 bags all together. Lucy filled twice as many bags as Marcus.

Lucy	Marcus	Total
12	6	18
14	7	21
16	8	24

I used **Guess and Test** and **Make a Chart**. I used numbers that were double one another until I found a sum of 24. The chart helped organize my guesses. Lucy filled 16 bags. Marcus filled 8 bags.

Quiz Problems

Here are some problems for you to try. Keep your **rubric** handy while you solve the problems. Let's see if you can score a **4**.

1. The basketball team can spend $200 for uniform shirts and basketballs. The coach bought 22 basketball shirts for $5 each and 9 basketballs for $8 each. Did the coach spend more or less than $200? How much more or less than $200 did she spend?

2. A cypress tree is 5 meters tall. An oak tree is 3 meters taller than the cypress tree. The palm tree is 2 meters taller than the oak tree. How tall is the palm tree?

3. Juan at the Cycle Shop sold 13 bicycles and tricycles today. The vehicles had a total of 33 wheels. How many of each did Juan sell?

4. Rory spent 3 times as many hours volunteering as Jordan. They spend a total of 24 hours a month volunteering. How many hours a month does each of them spend volunteering?

5. Michelle has 48 stamps from Peru. Fabian has 100 stamps from Peru. Rosemary has 50 stamps from Peru. They decided to share their Peruvian stamps equally. How many Peruvian stamps will they each have?

6. There are 60 actors, 18 dancers and 12 singers trying out for parts in the school play. The teacher chose 12 actors, $\frac{1}{2}$ as many dancers as actors, and 5 singers to be in the play. How many people were not picked to be in the play?

7. Adam bought 3 pounds of apples and 6 lemons. Apples sell for 59¢ a pound and lemons are 3 for 69¢. How much change will Adam receive from a $10.00 bill?

6. Algebra

In this chapter, we are going to look at some basic ideas of algebra. Algebra is about **discovering patterns** and **finding unknowns**. It makes you think about **number relationships** and **equations**, such as what is equal to what. You will also focus on your thoughts that lead to your best guess when solving problems.

Modeled Problem

Martha and Amy went to the bookstore to buy calendars. Martha spent **twice** as much as Amy. **Together** they spent $27. How **much** did each spend?

Keywords: twice, much, together

1. Read and Think

What **question** are you being asked?

● **How much did each girl spend?**

What are the **keywords**?

● **twice, much, together**

What are the **facts**?

● **They spent $27.**
● **Martha spent twice as much as Amy.**

2. Select a Strategy

We will use **Guess and Test** to solve the problem. We will also **Make a Table** to organize the information gathered as we Guess and Test.

3. Solve

We will pick pairs of numbers. One number must be twice the other number. Their sum must equal 27.

Martha	Amy	Total
10	5	15
12	6	18
14	7	21
16	8	24
18	9	27

Martha spent $18 and Amy spent $9.

4. Write/Explain

We had to find two addends that have a sum of $27. We needed to find one addend that was double the other. We used the **Guess and Test** strategy to find the numbers that matched these facts. The answers were Martha spent $18 and Amy spent $9.

5. Reflect

Let's review our work and answer.

- Did we show that we knew what the problem asked? **Yes. We answered the question that was asked. We found how much each girl spent.**

- Did we know what the keywords were? **Yes.**

- Did we show that we knew what facts were given? **Yes.**

- Did we name and use the correct strategy? **Yes.**

- Was our mathematics correct? **Yes. Our sum is equal to $27 and one addend is double the other.**

- Did we label our work? **Yes.**

- Was our answer correct? **Yes.**

- Were all of our steps included? **Yes.**

- Did we explain why we chose the strategy and how we used it? **Yes.**

- Did we write a good, clear explanation of our work? **Yes.**

Score

This solution would earn a perfect **4** on our rubric. The question asked, the keywords, and the facts were all given. Good strategies were used and the math was correct and labeled. The answer is explained well.

On the following pages are some **Guided Open-Ended Math Problems.**

For each problem there are **four parts**. In the **first part**, you will solve the problem with guided help. In the **second part**, you will score and correct a solution with guided help. The **third part** shows one solution that scores a perfect **4**. This solution may or may not differ from your way. The **fourth part** has *answers* to the **first** and **second parts** so you can check your work.

Guided Problem #1

Greg received 3 gifts. Uncle Pete's gift was $2.12 more than Aunt Sally's gift. Uncle Pete's gift cost $3.14 less than Uncle Ray's gift. Aunt Sally's gift cost $8.12. How much did the gifts cost all together?

Keywords: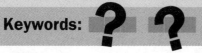

1. Try It Yourself.

Answer the questions below to get a score of **4**.

What **question** are you being asked?

_____

_____

What are the **keywords**?

_____

_____

What are the **facts** you need to solve the problem?

_____

_____

What **strategy** can you use to solve the problem?

_____

_____

> **Hint**
>
> Possible answers include: **Work Backward** and **Use Logical Thinking**.

Solve the problem.

_____

_____

_____

_____

Write/Explain what you did to solve the problem.

_____

_____

_____

Reflect. Review and improve your work.

_____

_____

2. Rick Tries It.

Rick's Paper

Question: How much did the gifts cost all together?

Keywords: more, less, much, all together

Facts: Uncle Pete's gift cost $2.12 more than Aunt Sally's gift.

Uncle Pete's gift cost $3.14 less than Uncle Ray's gift.

Aunt Sally's gift cost $8.12.

Strategy: Divide and Conquer

Solve: Let p = the amount of money Uncle Pete spent for his gift.

Let r = the amount of money Uncle Ray spent for his gift.

Let s = the amount of money Aunt Sally spent for her gift.

$s = 8.12$

$p = s - 2.12$, $p = 6$

$r = p + 3.14$, $r = \$9.14$

$\$8.12 + \$6.00 + \$9.14 = \$23.26$

The gifts cost $23.26 all together

Write/Explain: I used Divide and Conquer by writing number sentences. I determined what each relative spent. Then I added the amounts.

Score the Answer.

According to the rubric, from **1** to **3**, what score would you give Rick? Explain why you gave that score.

_____

_____

_____

_____

_____

_____

_____

Make it a 4! Rewrite.

_____

_____

_____

_____

_____

_____

_____

Use the rubric on *page 13* to score this work.

3. Maureen Tries It.

Remember, that there is often more than one way to solve a problem. Here is how Maureen solved this problem.

Maureen's Paper

Question: How much did the gifts cost all together?

Keywords: more, less, all together, much

Facts: Uncle Pete's gift was $2.12 more than Aunt Sally's gift.
Uncle Pete's gift was $3.14 less than Uncle Ray's gift.
Aunt Sally's gift cost $8.12.

Strategy: Work Backward

Solve:

Let r = the amount of money Uncle Ray spent for his gift

Let s = the amount of money Aunt Sally spent for her gift

Let p = the amount of money Uncle Pete spent for his gift.

$s = \$8.12$

$p = s + \$2.12 = \$8.12 + \$2.12 = \$10.24$

$r = p + \$3.14 = \$10.24 + \$3.14 = \$13.38$

$\$8.12 + \$10.24 + \$13.38 = \$31.74$

The gifts cost $31.74 all together—

Write/Explain: I used the Work Backward strategy. I used the information that I knew, which was the cost of Aunt Sally's gift, to find the cost of the other 2 gifts. Then I added to find the total amount spent.

Score: Maureen's paper would receive a **4** on the test. Maureen identified the question that was asked, the keywords, and the facts. She picked a good strategy and explained how she used it. Then she clearly explained the steps she took to solve the problem and labeled her work.

4. Answers to Parts 1 and 2.

Guided Problem #1

Greg received 3 gifts. Uncle Pete's gift was $2.12 more than Aunt Sally's gift. Uncle Pete's gift cost $3.14 less than Uncle Ray's gift. Aunt Sally's gift cost $8.12. How much did the gifts cost all together?

Keywords: less, more, all together

1. Try It Yourself. (page 49)

Question: What is the cost of the 3 gifts all together?

Facts: Uncle Pete's gift was $2.12 more than Aunt Sally's gift.

Uncle Pete's gift cost $3.14 less than Uncle Ray's gift.

Aunt Sally's gift cost $8.12.

Strategies: Work Backward, Write a Number Sentence

Solve: Aunt Sally's gift cost $8.12.

Uncle Pete's gift cost $2.12 more than Aunt Sally's gift: $8.12 + $2.12 = $10.24

Uncle Peter gift cost $3.14 less than Uncle Ray's gift: $10.24 + $3.14 = $13.38

$8.12 + $10.24 + $13.38 = $31.74

Write/Explain: I used the Work Backward and Write a Number Sentence strategies. I used the cost of Aunt Sally's gift to find the costs of the other 2 gifts. Then I added the costs of all the gifts to find the total amount spent.

2. Rick Tries It. (page 50)

Score the Answer: I would give Rick a **2** on his paper. He knew what question was asked, and gave the keywords, and the facts. He used good strategies, but he did not know what to do with his information. His number sentences did not follow the question. His explanation was not detailed.

Make it a 4! Rewrite.

$s = \$8.12$

$p = s + \$2.12 = \$8.12 + \$2.12 = \$10.24$

$r = p + \$3.14 = \$10.24 + \$3.14 = \$13.38$

$\$8.12 + \$10.24 + \$13.38 = \$31.74$

I used the Work Backward strategy. I used the cost of Aunt Sally's gift to find the cost of the other 2 gifts. Then I added to find the total amount spent on the gifts.

Guided Problem #2

Theo was born 7 days before Isabelle. Theo gave Isabelle the following number puzzle.

I am thinking of 2 numbers. When you add them together you get 42.

Their difference is 8. What are my 2 numbers?

Keywords:

1. Try It Yourself.

Answer the questions below to get a score of **4**.

What **question** are you being asked?

_____

_____

What are the **keywords**?

_____

_____

What are the **facts** you need to solve the problem?

_____

_____

What **strategy** can you use to solve the problem?

_____

_____

Solve the problem.

Hint

Possible Answers include: **Guess and Test** and **Write a Number Sentence**.

_____

_____

_____

_____

_____

Write/Explain what you did to solve the problem.

_____

_____

_____

_____

Reflect. Review and improve your work.

_____

_____

2. Helene Tries It.

Helene's Paper

Question: What are the 2 numbers?

Keywords: add, difference

Facts: 2 numbers add to 42.

They have a difference of 8.

Strategy: Write a Number Sentence

Solve:

Try 42 and 1	42 × 1 = 42	42 – 1 = 41
Try 21 and 2	21 × 2 = 42	21 – 2 = 19
Try 14 and 3	14 × 3 = 42	14 – 3 = 11
Try 7 and 6	7 × 6 = 42	7 – 6 = 1

There is no answer to the puzzle.

Write/Explain: I wrote a pair of number sentences. I have tried all the possible numbers. No other pair of factors multiply to 42. There were only 4 pairs of numbers.

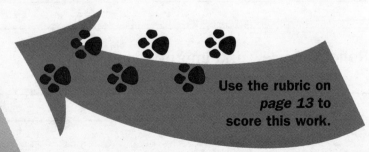

Use the rubric on *page 13* to score this work.

Score the Answer. According to the rubric, from **1** to **3**, what score would you give Helene? Explain why you gave that score.

_____

_____

_____

Make it a 4! Rewrite.

_____

_____

_____

_____

3. Ian Tries It.

Remember, there is often more than one way to solve a problem. On the following page, see how Ian solved this problem.

Ian's Paper

Question: What are 2 numbers that have a sum of 42 and a difference of 8?

Keywords: add, difference

Facts: 2 numbers add to 42.
The difference of the same 2 numbers is 8.

Strategy: I used Logical Thinking.

Solve: 42 ÷ 2 = 21
Since the difference is 8, add 4 to one side and subtract 4 from the other:

21 + 4 = 25   21 – 4 = 17
25 + 17 = 42   25 – 17 = 8

Write/Explain: I used Logical Thinking. The first fact, Theo being born 7 days before Isabelle, does not help solve the problem. I found half of 42, which is 21. Since the difference is 8, I added half of 8 to 21 and subtracted half of 8 from 21. I got the numbers 25 and 17, which have a sum of 42 and a difference of 8.

Score: Ian would receive a perfect 4. He gave the question, the keywords, and the facts, and answered the question. He used a good strategy, explained his work clearly, and labeled his answer.

4. Answers to Parts 1 and 2.

Guided Problem #2

Theo was born 7 days before Isabelle. Theo gave Isabelle the following number puzzle.

I am thinking of 2 numbers. When you add them together you get 42.

Their difference is 8. What are my 2 numbers?

Keywords: add, difference

1. Try It Yourself. (page 53)

Question: What are the 2 numbers that have a sum of 42 and a difference of 8?

Facts: 2 numbers add to 42.

The same 2 numbers have a difference of 8.

Strategy: Guess and Test

Solve:

| Try 24 and 18 | 24 + 18 = 42 | 24 – 18 = 6 |
| Try 25 and 17 | 25 + 17 = 42 | 25 – 17 = 8 |

Write/Explain: I used Guess and Test. I picked pairs of numbers that add to 42 until finding a pair that has a difference of 8. Those numbers are 25 and 17.

2. Helene Tries It. (page 54)

Score the Answer: I would give Helene a **2**. She knew the question, gave the keywords, used only the facts that were needed to solve the problem, and listed the facts. She picked a good strategy, but she gave up on the problem. She confused multiplication with addition. Her answer was incorrect.

Make it a 4! Rewrite.

Try 24 and 18	24 + 18 = 42	24 – 18 = 6
Try 25 and 17	25 + 17 = 42	25 – 17 = 8

I used Guess and Test. I picked pairs of numbers that add to 42 until finding a pair that has a difference of 8. Those numbers are 25 and 17. I ignored that Theo was born 7 days before Isabelle because it did not help solve the problem.

Guided Problem #3

Sandra is working with this function table.

Input	Output
2	5
5	11
7	15
9	?

What is the missing number?

Keywords:

1. Try It Yourself.

Answer the questions below to get a score of **4**.

What **question** are you being asked?

_____

What are the **keywords**?

_____

_____

What are the **facts** you need to solve the problem?

_____

_____

_____

_____

What **strategy** can you use to solve the problem?

_____

_____

Solve the problem.

_____

_____

_____

_____

Write/Explain what you did to solve the problem.

_____

_____

_____

Reflect. Review and improve your work.

Hint

Answers include **Look for a Pattern** and **Write a Number Sentence**.

2. Lars Tries It.

Lars's Paper

Question: What is the Output when the Input is 9?

Keywords: function, missing

Facts: 2 becomes 5, 5 becomes 11, 7 becomes 15.

Strategy: Look for a Pattern.

Solve: The difference in the Output between when 2 and 5 were Inputs was 6, so 15 + 6 = 21. The missing number is 21.

Write/Explain: I used the Look for a Pattern strategy. I determined that the difference between the Outputs for Inputs 2 and 5 was 6. So I added 6 to the Output of 15 to get 21.

Score the Answer. According to the rubric, from **1** to **3**, what score would you give Lars? Explain why you gave that score.

_____

_____

_____

_____

_____

Make it a 4! Rewrite.

_____

_____

_____

_____

_____

Use the rubric on *page 13* to score this work.

3. Patrick Tries It.

Remember, there is often more than one way to solve a problem. Here is how Patrick solved this problem.

Patrick's Paper

Question: What is the Output when the Input is 9?

Keywords: function, missing

Facts: 2 becomes 5, 5 becomes 11, and 7 becomes 15.

Strategy: I wrote an equation.

Solve: The equation Output = (2 × Input) + 1 is the rule of the function table.

$$(2 \times 9) + 1 = 19$$

The missing number is 19.

Write/Explain: I wrote the equation Output = (2 × Input) + 1 as the rule of the table. I then substituted 9 for the Input to get (2 × 9) + 1 = 19. The missing number is 19.

Score: Patrick's paper would earn a **4** on our test. He showed that he understood the question that was asked. He identified the keywords and the facts. He picked a good strategy. Then he clearly explained the steps he took to solve the problem and labeled his answer.

4. Answers to Parts 1 and 2.

Guided Problem #3

Sandra is working with this function table.

Input	Output
2	5
5	11
7	15
9	?

What is the missing number?

Keywords: function, missing

1. Try It Yourself. (page 56)

Question: What is the Output when the Input is 9?

Facts: 2 becomes 5, 5 becomes 11, 7 becomes 15

Strategy: Look for a Pattern

Solve: I Found a Pattern. The Output is always 1 more than double the Input.

$9 \times 2 = 18$, $18 + 1 = 19$. The missing number is 19.

Write/Explain: I found a pattern of the Output being 1 more than 2 times the Input. I doubled 9, which is 18, and then I added 1 to get 19. The missing number is 19.

Lars Tries It. (page 57)

I would give Lars a **3**. He knew the question, gave the keywords and the facts, and picked a good strategy. He made a mistake in his pattern, which caused him to get an incorrect answer. His explanation of what he did was good.

Make it a 4! Rewrite.

The pattern is the Output is 1 more than 2 times the Input. Multiply $9 \times 2 = 18$ and then add, $18 + 1 = 19$ to get the missing number.

Quiz Problems

Here are some problems for you to try. Keep your **rubric** handy while you solve the problems. Let's see if you can score a **4**.

1. Lorraine bought 2 notebooks and a new book bag. She spent $24. The 2 notebooks were the same price. The book bag cost twice as much as 1 of the notebooks. How much did she spend for each item?

2. Carlos sells worms to fishermen. On the first day he sold 12 worms. On the second day he sold 10 worms. On the third day he sold 15 worms. On the fourth day he sold 13 worms. On the fifth day he sold 18 worms. If the pattern continues, how many worms did Carlos sell on the seventh day?

3. In a hot-dog-eating contest, Manny ate 3 times as many hot dogs as Ollie. Stan ate 5 hot dogs fewer than Manny. Ollie ate 12 hot dogs. How many hot dogs did they eat in all?

4. Here is the puzzle that Joanie is working on.

There are 2 numbers. When you add them you get 18. When you find their difference, it is 4. What are the 2 numbers?

5. In this money function machine, you put money in and you get money out. You always get more money out than you put in.

Money In	Money Out
$1	$12
$2	$14
$3	$16
$4	$18

Find out how much money you get from this machine when $10 is put in.

6. Becky's dad sponsored her in a charity bike ride. He paid $0.01 for the first $\frac{1}{2}$ mile, $0.02 for the second $\frac{1}{2}$ mile, $0.04 for the third $\frac{1}{2}$ mile, and $0.08 for the next $\frac{1}{2}$ mile, with the pattern continuing for the distance she biked. Becky rode 7 miles. How much did her dad have to pay for the last $\frac{1}{2}$ mile?

7. How many dots are in the sixth part of this diagram?

7. Geometry

Look around your home, your school, your neighborhood. What **shapes** do you see? Squares? Cylinders? Triangles? What **lines** do you see? Some that never meet? Some that intersect? Some shorter? Some longer? What **angles** do you see? Some that are right angles? Some that are not? You are looking at geometry. Geometry is part of mathematics. Geometry is part of your everyday life.

Here is a geometry problem that might be on your test. Let's see how we can solve this problem and get a perfect score of **4** on our rubric.

Modeled Problem

What **best** describes the **3-dimensional figure** that can be **constructed** from this **net**?

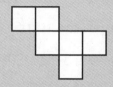

Keywords: best, 3-dimensional figure, constructed, net

1. Read and Think

What **question** are we asked?

- **What 3-dimensional figure best describes the net?**

What are the **keywords**?

- **best, 3-dimensional figure, constructed, net**

What **facts** are we given?

- **We know the 3-dimensional figure has 6 faces.**

2. Select a Strategy

We can use **Logical Thinking**.

3. Solve

What I know:

- **The figure has 6 faces.**
  A rectangular prism and a cube have 6 faces.
- **The faces are all squares.**
  A cube has 6 square faces.
- **The figure is a cube.**

4. Write/Explain

We use what we know from the drawing and apply **Logical Thinking** to solve the problem. We can see that the net has 6 faces. Rectangular prisms and cube have 6 faces. A rectangular prism is made of 6 rectangles, but a cube must be made of 6 squares. Cube best describes the figure.

5. Reflect

Let's review our work and answer.

- Did we show that we knew what the problem asked? **Yes. We answered the question that was asked.**

- Did we know what the keywords were? **Yes.**

- Did we show that we knew what facts were given? **Yes.**

- Did we name and use the correct strategy? **Yes.**

- Was our mathematics correct? **Yes. We checked it. It was correct.**

- Did we label our work? **Yes.**

- Was our answer correct? **Yes.**

- Were all of our steps included? **Yes.**

- Did we explain why we chose the strategy and how we used it? **Yes.**

- Did we write a good, clear explanation of our work? **Yes.**

On the following pages are some **Guided Open-Ended Math Problems**.

For each problem there are **four parts**. In the **first part**, you will solve the problem with guided help. In the **second part**, you will score and correct a solution with guided help. The **third part** shows one solution that scores a perfect **4**. This solution may or may not differ from your way. The **fourth part** has *answers* to the **first** and **second parts** so you can check your work.

Use the rubric on *page 13* to score this work.

Guided Problem #1

Rosemary is going to design a figure on a coordinate grid. She has written ordered pairs at (1, 1), (1, 3), and (3, 3). What shape did she make? Answer as specifically as you can.

Keywords:

1. Try It Yourself.

Answer the questions below to get a score of **4**.

What **question** are you being asked?

_____

What are the **keywords**?

_____

_____

What are the **facts** you need to solve the problem?

_____

_____

_____

_____

What **strategy** can you use to solve the problem?

_____

_____

Hint

Possible answers include: **Draw a Picture**, **Logical Thinking**.

Solve the problem.

_____

_____

_____

_____

Write/Explain what you did to solve the problem.

_____

_____

Reflect. Review and improve your work.

_____

_____

Patti Tries It.

Patti's Paper

Question: What shape can be made from the ordered pairs?

Keywords: figure, coordinate grid, ordered pairs, shape, specifically

Facts: There are ordered pairs at (1, 1), (1, 3), and (3, 3).

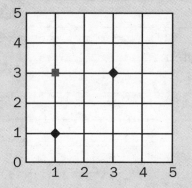

Write/Explain: I Drew a Picture. I plotted the ordered pairs on a coordinate grid. Then I connected the points. The figure that I created has 3 sides, so it is a triangle.

Score the Answer. According to the rubric, from **1** to **3**, what score would you give Patti? Explain why you gave that score.

_____

_____

_____

_____

_____

_____

Make it a 4! Rewrite.

_____

_____

_____

_____

_____

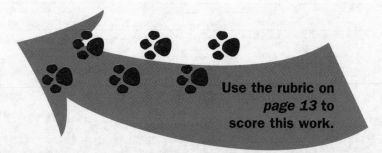

Use the rubric on *page 13* to score this work.

Remember, there is often more than one way to solve a problem. Here is how Chris solved this problem.

Chris's Paper

Question: What shape can be made from the ordered pairs?

Keywords: figure, coordinate grid, ordered pairs, shape, specifically

Facts: There are ordered pairs at $(1, 1)$, $(1, 3)$, $(3, 3)$.

Strategy: I used Logical Thinking.

Solve: There are 3 ordered pairs. A figure with 3 sides is a triangle.

If the triangle was drawn on a grid, there would be perpendicular segments. This means it is a right triangle.

The distance between the first coordinates is 2.

The distance between the second coordinates is 2.

Two of the sides are equal. The third side is longer. This means it is an isosceles right triangle.

Write/Explain: I used Logical Thinking. I counted the number of points and determined the figure was a triangle. The ordered pairs form perpendicular segments, so the triangle is a right triangle. I subtracted the difference between the coordinates to find the length of the sides. The lengths of 2 of the sides are equal. The third side is longer than the other two sides. So the figure is an isosceles right triangle.

Score: Chris's solution would earn a **4** on a test. Chris identified the question that was asked, the keywords, and the facts. He picked a good strategy. Chris clearly explained the steps taken to solve the problem and labeled his work. Good job!

4. Answers to Parts 1 and 2.

Guided Problem #1

Rosemary is going to design a figure on a coordinate grid. She has written ordered pairs at (1, 1), (1, 3), and (3, 3). What shape did she make? Answer as specifically as you can.

Keywords: figure, coordinate grid, ordered pairs, shape, specifically

1. Try It Yourself. (page 65)

Question: What shape can be made?

Facts: Points are at (1, 1), (1, 3), and (3, 3).

Strategy: Draw a Picture.

Solve:

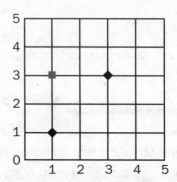

When I connect the ordered pairs, I find that the figure is a triangle with 2 sides the same length. It also has a right angle, so the figure is an isosceles right triangle.

Write/Explain: I **Drew a Picture**. I plotted the ordered pairs on a coordinate grid. Then I connected the points. The figure that I created was a triangle with 2 equal sides and a right angle, so it is an isosceles right triangle.

2. Patti Tries It. (page 66)

Score the Answer: I would give Patti a **3**. She knew what question was being asked, gave the keywords, and listed the facts. Her answer was not wrong. However, she did not answer the question as specifically as she could have.

Make it a 4! Rewrite.

Everything other than giving the specific triangle was correct. The figure has a right angle and 2 equal sides, so it is an isosceles right triangle.

Guided Problem #2

Andy said that all rectangles are similar. Reggie said that Andy was incorrect and that all squares are similar. Who is correct? Explain why.

Keywords:

1. Try It Yourself.

Answer the questions below to get a score of **4**.

What **questions** are you being asked?

_____

_____

What are the **keywords**?

_____

_____

_____

What are the **facts** you need to solve the problem?

_____

_____

What **strategy** can you use to solve the problem?

_____

_____

Hint

Possible answers include: **Draw a Picture** and **Logical Thinking**.

Solve the problem.

_____

_____

_____

_____

Write/Explain what you did to solve the problem.

_____

_____

Reflect. Review and improve your work.

_____

_____

_____

2. Joy Tries It.

Joy's Paper

Questions: Are all rectangles similar?
Are all squares similar?

Keywords: rectangles, similar, squares

Facts: Reggie and Andy disagree on whether rectangles are similar. Instead, Andy thinks all squares are similar.

Strategy: Draw a Picture.

Solve: I made a drawing of 2 squares and 2 rectangles.

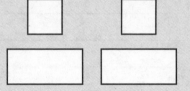

They are both right. All rectangles and squares are similar.

Write/Explain: I Drew a Picture of 2 squares. They are both similar. I Drew a Picture of 2 rectangles. They are both similar. So, both Andy and Reggie were correct since all rectangles are similar and all squares are similar.

Score the Answer.

According to the rubric, from **1** to **3**, what score would you give Joy? Explain why you gave that score.

_____

_____

_____

Make it a 4! Rewrite.

_____

_____

_____

3. Dominic Tries It.

Remember, there is often more than one way to solve a problem. Here is how Dominic solved this problem.

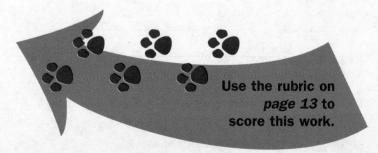

Use the rubric on *page 13* to score this work.

Dominic's Paper

Questions: Are all rectangles similar to each other? Are all squares similar to each other?

Keywords: rectangles, similar, squares

Facts: Reggie and Andy disagree. Reggie thinks all rectangles are similar. Andy thinks all squares are similar.

Strategy: I used Logical Thinking.

Solve: Both rectangles and squares are parallelograms with 4 right angles. A rectangle must have opposite sides equal and a square must have all sides equal. Therefore, a rectangle can have many shapes and a square can have only 1 shape. Reggie is correct.

Write/Explain: I used Logical Thinking. I listed the characteristics of rectangles and squares. I wrote the characteristics that they share and the ones where they differ. Some rectangles are similar but not all rectangles. All squares are similar. Reggie is correct.

Score: Dominic's solution would earn a **4** on a test. He identified the question that was asked, the keywords, and the facts, and picked a good strategy. He gave an example of how his strategy works and explained what he did. He labeled his answer.

4. Answers to Parts 1 and 2.

Guided Problem #2

Andy said that all rectangles are similar. Reggie said that Andy was incorrect and that all squares are similar. Who is correct? Explain why.

Keywords: rectangles, similar, squares

1. Try It Yourself. (page 69)

Questions: Are all rectangles similar? Are all squares similar?

Facts: Andy thinks all rectangles are all similar. Reggie disagrees and thinks all squares are similar.

Strategy: Draw a Picture.

Are all squares similar?

Solve:

All squares are similar, so Reggie is correct.

Write/Explain: I Drew a Picture of 2 rectangles that are not similar. I then drew 3 squares and all squares were similar to each other. I could have drawn more, but they would all have the same shape. Therefore, Reggie is correct because all squares are similar, but not all rectangles are similar.

2. Joy Tries It. (page 70)

Score the Answer: I would give Joy a **2**. She knew what question was being asked, gave the keywords, and listed the facts. Her answer was incorrect. She didn't know that rectangles can have sides of different lengths. So rectangles are not all similar.

Check the Glossary on *p. 139*

Make it a 4! Rewrite.

Joy needs to try to draw figures that are not similar.

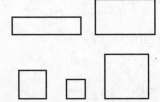

The rectangles are not similar. The squares are all similar. Therefore, Reggie is correct because all squares are similar to each other, but not all rectangles are similar to each other.

Guided Problem #3

What are the angle classifications for each type of triangle? Explain your answer.

Keywords:

1. Try It Yourself.

Answer the questions below to get a score of **4**.

What **question** are you being asked?

_____

What are the **keywords**?

_____

_____

What are the **facts** you need to solve the problem?

_____

What **strategy** can you use to solve the problem?

_____

_____

Solve the problem.

_____

_____

_____

_____

Write/Explain what you did to solve the problem.

_____

_____

_____

_____

_____

Reflect. Review and improve your work.

_____

_____

_____

_____

2. Kenny Tries It.

Kenny's Paper

Question: Which angles make up each kind of triangle?

Keywords: classifications, each, triangle

Facts: A triangle is made of 3 angles.

Solve:

3 angles   3 angles   3 angles

(continue working on Kenny's problem on the following page.)

Use the rubric on *page 13* to score this work.

Score the Answer.

According to the rubric, from **1** to **3** what score would you give Kenny? Explain why you gave that score.

_____

_____

_____

_____

_____

Make it a 4! Rewrite.

_____

_____

_____

_____

3. Anjali Tries It.

Remember, there is often more than one way to solve a problem. Here is how Anjali solved this problem.

Anjali's Paper

Question: What are the angle classifications for each type of triangle?

Keywords: classifications, each, triangle

Facts: A triangle has 3 angles.

Strategy: I Made a Table.

Solve:

Acute Triangle	Right Triangle	Obtuse Triangle
3 acute angles	2 acute angles	2 acute angles
	1 right angle	1 obtuse angle

Write/Explain: I Made a Table. I know that a triangle can have only 1 right or obtuse angle. Therefore, the other two angles must be acute. An acute triangle has only acute angles.

Score: Anjali would earn a **4** on our rubric. She identified the question that was asked, the keywords, and the facts, and picked a good strategy. Her explanation was clear and complete, and she labeled her answer.

4. Answers to Parts 1 and 2.

Guided Problem #3

What are the angle classifications for each type of triangle? Explain your answer.

Keywords: classifications, each, triangle

1. Try It Yourself. (pages 72–73)

Question: What are the angle classifications for each type of triangle?

Facts: A triangle is made of 3 angles.

Strategy: Use Logical Thinking.

Solve: A triangle has 3 angles. Only 1 of those angles can be a right or obtuse angle. Therefore, every triangle must have at least 2 acute angles.

Acute triangle: 3 acute angles.

Right triangle: 2 acute angles, 1 right angle.

Obtuse triangle: 2 acute angles, 1 obtuse angle.

Write/Explain: A triangle can have only 1 right angle or 1 obtuse angle. So, every triangle must have at least 2 acute angles. An acute triangle will have 3 acute angles and the right and obtuse triangles will have 2 acute angles. A right triangle will have 1 right angle and an obtuse triangle will have 1 obtuse angle.

2. Kenny Tries It. (pages 73–74)

Score the Answer: I would give Kenny a **2**. He gave the question, the keywords, and the facts. He drew all three types of triangles. However, he did not identify those angles. He also did not write an explanation of his work.

Make it a 4! Rewrite.

Acute Triangle	Right Triangle	Obtuse Triangle
3 acute angles	2 acute angles	2 acute angles
	1 right angle	1 obtuse angle

I Drew a Picture of each type of triangle. An angle less than a right angle is an acute angle. An angle greater than a right angle is an obtuse angle. Beneath each drawing (on page 73) is a description of the 3 angles.

Quiz Problems

Here are some problems for you to try. Keep your **rubric** handy while you solve the problem. Let's see if you can score a **4**.

1. Two triangles are needed to make a square. How many triangles are needed to make a hexagon? Explain your answer.

2. Can Samantha draw a parallelogram with a base of 6 inches and a height of 4 inches, and a rectangle with a length of 6 inches and a height of 4 inches that are not congruent?

3. Which of the letters in **HARK** has rotational symmetry? Explain how you know.

4. How are a chord and a diameter alike? How are they different?

5. Debra has plotted an ordered pair at (3, 2). If she moves the ordered pair 2 units left and 3 units up, where did she plot the new ordered pair?

6. What 3-dimensional figure does this net make?

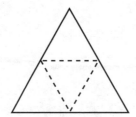

7. How many squares can you find in this diagram? *Hint:* The answer is not 16.

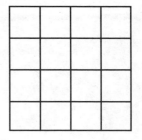

8. Measurement

There are all kinds of measurements on tests. You have to know about units of **length**, which include **inches, feet, yards, centimeters,** and **meters**. You have to know about units of **capacity**, such as **cups, pints, quarts, gallons,** and **liters**. You have to know about units like **ounces, pounds, grams,** and **kilograms** that are used to measure **weight** and **mass**.

Money, time and temperature are also kinds of measurements. Let's look at a problem in measurement that you might find on a test. Let's see how to solve it. Remember, we want to get a perfect score of **4** on our scoring rubric.

Modeled Problem

Paul is framing a picture with red yarn. The picture is **rectangular** and is 30 inches long and 24 inches wide. How **much** red yarn does Paul need to make the frame?

Keywords: rectangular, much

1. Read and Think

What **question** are we asked?

● **What is the perimeter of the frame?**

What are the **keywords**?

● rectangular
● much

What are the **facts**?

● **The frame will be 30 in. long and 24 in. wide.**

2. Select a Strategy

We can **Draw a Picture** to see what is happening in the problem.

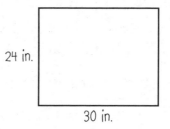

24 in.

30 in.

3. Solve

The distance around a figure is its perimeter. In a rectangle, the opposite sides are equal. We can add the lengths of the sides to find the perimeter.

30 in. + 24 in. + 30 in. + 24 in. = 108 in.

The perimeter is 108 inches.

Paul needs 108 inches of yarn to make the frame.

4. Write

To solve this problem, I **Drew a Picture**. I had to find the perimeter of the rectangle. I knew the opposite sides were equal, so I knew the length of all four sides. I added the sides to find that the perimeter is 108 inches.

5. Reflect

Let's review our work and answer.

- Did we show that we knew what the problem asked? **Yes. We answered the question that was asked.**

- Did we know what the keywords were? **Yes.**

- Did we show that we knew what facts were given? **Yes.**

- Did we name and use the correct strategy? **Yes.**

- Was our mathematics correct? **Yes. We checked it. It was correct.**

- Did we label our work? **Yes.**

- Was our answer correct? **Yes.**

- Were all of our steps included? **Yes.**

- Did we explain why we chose the strategy and how we used it? **Yes.**

- Did we write a good, clear explanation of our work? **Yes.**

Score

This would earn a **4** on our rubric. All the work is correct. The strategy was a good one.

On the following pages are some

Guided Open-Ended Math Problems.

For each problem there are **four parts.** In the **first part**, you will solve the problem with guided help. In the **second part**, you will score and correct a solution with guided help. The **third part** shows one solution that scores a perfect **4**. This solution may or may not differ from your way. The **fourth part** has *answers* to the **first** and **second parts** so you can check your work.

Guided Problem #1

Here is a table showing the visiting hours for three museums.

Museum Visiting Hours	
Modern Art Museum	10:00 a.m. to 5:30 p.m.
Science Museum	9:30 a.m. to 4:45 p.m.
Children's Museum	9:30 a.m. to 5:00 p.m.

At 2:45 p.m., Mrs. Reynolds put enough money in the parking meter to last 2 hours 30 minutes. Will the money in the meter expire before the Children's Museum closes?

Keywords:

1. Try It Yourself.

Answer the questions below to get a score of **4**.

What **question** are you being asked?

_____

What are the **keywords**?

_____

_____

What are the **facts** you need to solve the problem?

_____

_____

What **strategy** can you use to solve the problem?

_____

_____

Solve the problem.

Hint

Possible answers include **Write a Number Sentence**, **Draw a Picture**, and **Divide and Conquer**.

_____

_____

Write/Explain what you did to solve the problem.

_____

_____

Reflect. Review and improve your work.

_____

2. Edie Tries It.

Edie's Paper

Question: Will the money last until 5:00 p.m.?

Keywords: enough, expire

Facts: The money went into the meter at 2:45. There is enough money for 2 hours 30 minutes.

Strategy: Wrote Number Sentences.

Solve:
$$
\begin{array}{r}
2{:}45 \\
+\ 2{:}30 \\
\hline
4{:}75
\end{array}
$$

$4{:}75 = 4\frac{3}{4}$ hour. The meter will not last until the museum closes at 5:00.

Write/Explain: I added the amount of time on the meter (2:45) to the amount of time Mrs. Reynolds put in (2:30). I got 4:75. Since 4.75 is equal to $4\frac{3}{4}$, the money in the meter will last until 4:45. The museum closes at 5:00, so there is not enough time on the meter.

Score the Answer. According to the rubric from **1** to **3**, what score would you give Edie? Explain why you gave that score.

_____

_____

_____

_____

_____

_____

Make it a 4! Rewrite.

_____

_____

_____

_____

_____

Use the rubric on *page 13* to score this work.

3. Katrina Tries It.

Remember, there is often more than one way to solve a problem. Here is how Katrina solved this problem.

Katrina's Paper

Question: Did she put enough money in to last until 5:00 p.m.?

Keywords: enough, expire

Facts: Mrs. Reynolds put money in the meter at 2:45 p.m.
She put enough money to last 2 hour 30 minutes.

Strategy: I Made 2 Drawings.

Solve:

Write/Explain: I Drew a Picture of the time 2 hours after 2:45, which is 4:45. I then made a drawing of 30 minutes after 4:45, which is 5:15. The money will last until 5:15, so Mrs. Reynolds put enough money in the meter to last until the museum closes at 5:00.

Score: Katrina's paper would earn a **4** on our rubric. She identified the question that was asked. She showed the keywords. Katrina selected a good method for solving the problem. Her work was correct. She wrote a paragraph that was clear and explained what she had done. She labeled her answer.

4. Answers to Parts 1 and 2.

Guided Problem #1

Here is a table showing the visiting hours for three museums.

Museum Visiting Hours	
Modern Art Museum	10:00 a.m. to 5:30 p.m.
Science Museum	9:30 a.m. to 4:45 p.m.
Children's Museum	9:30 a.m. to 5:00 p.m.

At 2:45 p.m., Mrs. Reynolds put enough money in the parking meter to last 2 hours 30 minutes. Will the money in the meter expire before the Children's Museum closes?

Keywords: enough, expire

1. Try It Yourself. (pages 81–82)

Question: Is there enough money for the meter to go past 5 p.m.?

Facts: Mrs. Reynolds put the money in the meter at 2:45 p.m.

She put enough money for 2 hours 30 minutes in the meter.

Strategy: Divide and Conquer.

Solve: 2:45 to 3:00 is 15 minutes
3:00 to 5:00 is 2 hours
5:00 to 5:15 is 15 minutes

The meter will expire at 5:15. There is enough money in the meter.

Write/Explain: I counted the minutes until the next hour. Then I counted 2 hours. Finally, I counted the remaining minutes. The time went beyond 5:00, so there is enough money in the meter.

2. Edie Tries It. (page 82)

Score the Answer: I would give Edie a **2** on this question. She knew what question was being asked. She gave the keywords and the facts. She chose a good strategy, but did not have the correct answer. She made the mistake of thinking that 75 minutes = $\frac{3}{4}$ hour instead of $1\frac{1}{4}$ hours.

Make it a 4! Rewrite.

Because there are 60 minutes in an hour, 4 hours 75 minutes = 5 hours 15 minutes. The meter had enough money until 5:15, so Mrs. Reynolds put enough money in the meter.

Guided Problem #2

Ruthie brought 3 quarts of orange juice to a class picnic. Melissa brought 5 pints of grape juice and Ling brought 72 fluid ounces of apple juice to the picnic. Who brought the most juice? Who brought the least juice?

Keywords: ? ?

1. Try It Yourself.

Answer the questions below to get a score of **4**.

What **questions** are you being asked?

_____

_____

What are the **keywords**?

_____

_____

_____

_____

What are the **facts** you need to solve the problem?

_____

_____

_____

What **strategy** can you use to solve the problem?

_____

_____

Solve the problem.

_____

_____

_____

Hint

Possible answers include: **Divide and Conquer**, **Draw a Picture**, and **Write a Number Sentence**.

Write/Explain what you did to solve the problem.

_____

_____

Reflect. Review and improve your work.

_____

2. Suzanne Tries It.

Suzanne's Paper

Questions: Who brought the most juice? Who brought the least juice?

Keywords: quarts, pints, fluid ounces, most, least

Facts: Ruthie brought 3 quarts. Melissa brought 5 pints. Ling brought 72 fluid ounces.

Strategy: Divide and Conquer.

Solve: 3 quarts × 2 = 6 pints
 5 pints = 5 pints
 72 ÷ 12 = 6, so 72 fluid ounces = 6 pints.

Ling and Ruthie each brought the most. Melissa brought the least.

Write/Explain: I used the Divide and Conquer strategy. I know that 3 quarts are equal to 6 pints. I found the number of pints that are equal to 72 fluid ounces. Then I compared the answers.

Score the Answer. According to the rubric, from **1** to **3**, what score would you give Suzanne? Explain why you gave that score.

_____

_____

_____

_____

_____

Make it a 4! Rewrite.

_____

_____

_____

_____

_____

_____

Use the rubric on _page 13_ to score this work.

3. Kyle Tries It.

Remember, there is often more than one way to solve a problem. Here is how Kyle solved this problem.

Kyle's Paper

Questions: Who brought the most juice? Who brought the least juice?

Keywords: quarts, pints, fluid ounces, most, least

Facts: Ruthie brought 3 quarts. Melissa brought 5 pints.
Ling brought 72 fluid ounces.

Strategy: I used Divide and Conquer.

Solve:
There are 32 fl oz in a quart, so $32 \times 3 = 96$ fluid ounces.
There are 16 fl oz in a pint, so $16 \times 5 = 80$ fluid ounces.

$$96 > 80 > 72$$

Since Ruthie brought the 3 quarts she brought the most juice.
Since Ling brought the 72 fluid ounces, she brought the least juice.

Write/Explain: I used the Divide and Conquer strategy. I calculated how many fluid ounces there were in 3 quarts and in 5 pints. Then I compared and ordered the number of fluid ounces to find who brought the most and least juice.

Score: Kyle would receive a **4** on our rubric. He answered the question that was asked. He understood the keywords. He used the given facts. He picked a good strategy and carried it out correctly. Then he explained his work carefully and clearly and labeled his work.

4. Answers to Parts 1 and 2.

Guided Problem #2

Ruthie brought 3 quarts of orange juice to a class picnic. Melissa brought 5 pints of grape juice and Ling brought 72 fluid ounces of apple juice to the picnic. Who brought the most juice? Who brought the least juice?

Keywords: quarts, pints, fluid ounces, most, least

1. Try It Yourself. (page 85)

Questions: Who brought the most juice? Who brought the least juice?

Facts: Ruthie brought 3 quarts. Melissa brought 5 pints. Ling brought 72 fluid ounces.

Strategy: Divide and Conquer.

Solve: 1 quart = 2 pints

3 quarts > 5 pints

$72 \div 32 = 2$ R8, so 3 quarts > 72 fluid ounces

Ruthie brought the most juice.

1 pint = 16 fluid ounces

$16 \times 5 = 80$ fluid ounces.

$80 > 72$

Ling brought the least juice.

Write/Explain: I used the **Divide and Conquer** strategy. *First* I compared the amounts that Ruthie and Melissa brought. Ruthie brought more. *Then* I compared the amounts that Ruthie and Ling brought. Ruthie brought more, so she brought the most. I compared the amounts that Melissa and Ling brought. Melissa brought more, so Ling brought the least.

2. Suzanne Tries It. (page 86)

Score the Answer: I would give Suzanne a **2**. She knew what the question asked, and gave the keywords and the facts. She answered the question incorrectly because she did not know that there are 16 fluid ounces in a pint.

Make it a 4! Rewrite.

3 quarts $\times$ 2 = 6 pints

5 pints = 5 pints

$72 \div 16 = 4$ R8, so 72 fluid ounces = 4 pints 1 cup.

Ruthie brought the most juice and Ling brought the least juice.

Guided Problem #3

Sarah is taller than Emily. Sarah and Emily each took a handful of coins.

Sarah	Emily
6 quarters	5 quarters
10 dimes	7 dimes
1 nickel	14 nickels
4 pennies	3 pennies

Who took more money? How much more money?

Keywords:

1. Try It Yourself.

Answer the questions below to get a score of **4**.

What **questions** are you being asked?

_____

_____

What are the **keywords**?

_____

What are the **facts** you need to solve the problem?

_____

_____

_____

_____

What **strategy** can you use to solve the problem?

_____

_____

Solve the problem.

_____

_____

_____

_____

_____

Write/Explain what you did to solve the problem.

_____

_____

_____

Reflect. Review and improve your work.

_____

> **Hint**
>
> Possible answers include **Divide and Conquer**, **Make a Table**, and **Draw a Picture**.

2. Johnny Tries It.

Johnny's Paper

Sarah has $1.50 + $1.00 + $0.05 + $0.04 = $2.59.

Emily has $1.25 + $0.70 + $0.70 + $0.03 = $2.68.

Emily has $2.68 – $2.59 = $0.09 more than Sarah.

Write/Explain: I found out how much money each girl had. Then I subtracted to find out how much more money Emily had. Then she should give that to Sarah and they will have the same amount.

Use the rubric on *page 13* to score this work.

Score the Answer. According to the rubric, scores range from **1** to **3**. What score would you give Johnny? Explain why you gave him that score.

_____

_____

_____

_____

_____

_____

Make it a 4! Rewrite.

_____

_____

_____

_____

_____

_____

3. Lisa Tries It.

Remember, there is often more than one way to solve a problem. Here is how Lisa solved this problem.

Lisa's Paper

Questions: Who took more money? How much money did the person who took more take than the person who took less?

Keywords: each, much, more

Facts: Sarah took 6 quarters, 10 dimes, 1 nickel, and 4 pennies.
Emily took 5 quarters, 7 dimes, 14 nickels, and 3 pennies.

Strategy: I Made a Table and I used Number Sentences.

Solve:

Sarah		Emily
6 × $0.25 = $1.50	Quarters	5 × $0.25 = $1.50
10 × $0.15 = $1.00	Dimes	7 × $0.15 = $0.70
1 × $0.05 = $0.05	Nickels	14 × $0.05 = $0.70
4 × $0.01 = $0.05	Pennies	3 × $0.01 = $0.03
$2.59	TOTAL	$2.68

Emily has 9¢ more than Sarah.

Write/Explain: I Made a Table to find how much money each girl had. I determined the value of each type of coin they had and added to find the total amount. Then I subtracted Sarah's amount from Emily's amount. $2.68 − $2.59 = $.09.

Score: Lisa would receive a 4 on our rubric. She answered the question that was asked. She used only the facts that were needed to solve the problem. She selected a good strategy, used the given information, and found the correct answer. She explained what she did very well. She labeled her answer.

4. Answers to Parts 1 and 2.

Guided Problem #3

Sarah is taller than Emily. Sarah and Emily each took a handful of coins.

Sarah	Emily
6 quarters	5 quarters
10 dimes	7 dimes
1 nickel	14 nickels
4 pennies	3 pennies

Who took more money? How much more money?

Keywords: each, much, more

1. Try It Yourself. (page 89)

Questions: Who took more money? How much more money did the person who took more take than the person who took less?

Facts: Sarah took 6 quarters, 10 dimes, 1 nickel, and 4 pennies.

Emily took 5 quarters, 7 dimes, 14 nickels, and 3 pennies.

Strategies: I used Divide and Conquer, Write Number Sentences.

Solve:

Sarah

$(6 \times \$0.25) + (10 \times \$0.10) + \$0.05 + \$0.04$

$\$1.50 + \$1.00 + \$0.05 + \$0.04 = \$2.59$

Emily

$(5 \times \$0.25) + (7 \times \$0.10) + (14 \times \$0.05) + \$0.03$

$\$1.25 + \$0.70 + \$0.70 + \$0.03 = \$2.68$

Emily took $\$2.68 - \$2.59 = \$0.09$ more than Sarah.

Write/Explain: I used Divide and Conquer to find how much each girl took. I did this by *first* writing a number sentence by multiplying the number of coins by the value of each coin. *Then* I added the products. I did not multiply when all I had to do was multiply by 1.

I determined that Emily took $0.09 more than Sarah.

2. Johnny Tries It. (page 90)

Score the Answer: I would give Johnny a **2** on this question. He did everything correctly, but he did not supply the question that he was given, the keywords, or the facts. He did use a good strategy and found the correct answer.

Make it a 4! Rewrite.

Questions: Who took more money? How much more money did the person who took more take than the person who took less?

Keywords: each, much, more

Facts: Sarah took 6 quarters, 10 dimes, 1 nickel, and 4 pennies.

Emily took 5 quarters, 7 dimes, 14 nickels, and 3 pennies.

Strategy: Divide and Conquer

All of Johnny's work was correct.

Quiz Problems

Here are some problems for you to try. Keep your **rubric** handy while you solve the problems. Let's see if you can score a **4**.

1. When Jan went to sleep there was a puddle in her yard. The temperature outside was 37°F. When she woke up, the puddle was frozen. What is the least number of °F the temperature could have fallen during the night?

2. Marti is 4 feet 3 inches tall. Her father is 27 inches taller than she. What is Marti's father's height, in feet and inches?

3. Nikki's bedroom is rectangular. It is 12 feet long by 9 feet wide. She is going to carpet her entire bedroom. How much carpeting does she need to buy?

4. Bill started studying at 10:45 a.m. He studied for 1 hour 30 minutes and then took a 20-minute break for lunch. He then studied until 3:15 p.m. How long did Bill study?

5. Katherine was 8 pounds 4 ounces when she was born. When Drew was born he weighed 104 ounces. Who weighed more at birth? How much more? Write your answer in pounds and ounces.

6. Lauren wants to construct a play area outside for her dog Truffles. She will put a fence on 3 sides of a 25-foot by 15-foot region. Lauren's house will supply the fourth boundary, which is one of the long ends. How much fencing does Lauren need to buy?

7. A fish tank is 4 feet long, 3 feet high, and 2 feet wide. What is the volume of the fish tank?

9. Data Analysis and Graphs

What would you do if you had a lot of **data** to analyze? How would you **make sense** of it? How would you make it worthwhile and not wasted? How would you organize it? The data should be organized in some way. You may choose to use **tables, line plots, bar graphs,** or **line graphs**. You should choose the way you think displays the data best. When you do so, you can study the data in a logical way, and use it to draw conclusions. Let's look at a model problem and see how it is done.

Modeled Problem

Mabel recorded the time she spent on her activities for a full day in a circle graph.

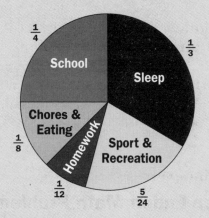

On which activity did Mabel spend the most time?

How many hours did she spend on that activity?

Keywords: circle graph, most, many

1. Read and Think

What **questions** are we asked?

● **On which activity did Mabel spend the most time?**

● How many hours did Mabel spend on the activity on which she spent the most time?

What are the **keywords**?

● **circle graph, most, many**

What **facts** are we given?

● The circle graph shows that Mabel was

  at school $\frac{1}{4}$ of the day

  slept $\frac{1}{3}$ of the day

  played sports $\frac{5}{24}$ of the day

  did homework for $\frac{1}{12}$ of the day

  did chores or ate $\frac{1}{8}$ of the day

  Full day is 24 hours.

2. Select a Strategy

We will **Read a Graph** and use **Logical Thinking** to answer the questions.

3. Solve

From the graph, we can see that Mabel spends more time sleeping than any other single activity. The graph says she spends $\frac{1}{3}$ of her day sleeping. To find $\frac{1}{3}$ of 24, divide 24 by 3. Since $24 \div 3 = 8$, Mabel spends 8 hours sleeping each day.

4. Write/Explain

To solve this problem, we read and interpreted a circle graph. The largest section of the circle graph was for sleep. I know there are 24 hours in each day. I found how many hours she slept. She spent 8 hours sleeping.

5. Reflect

Let's review our work and answer.

- Did we show that we knew what questions were asked? **Yes. We answered both of them.**

- Did we know what the keywords were? **Yes.**

- Did we show that we knew what facts were given? **Yes.**

- Did we name and use the correct strategy? **Yes. We named and used two strategies: Read a Graph and Logical Thinking**

- Was our mathematics correct? **Yes.**

- Did we label our work? **Yes. The answer was given in hours.**

  - Was our answer correct? **Yes.**

    - Were all of our steps included? **Yes.**

- Did we explain why we chose the strategy and how we used it? **Yes.**

- Did we write a good, clear explanation of our work? **Yes.**

Score

This solution would earn a perfect **4** on our rubric. We answered both questions that were asked. The strategies we chose were applied correctly. The answers we got are correct. The explanation was clearly written and made sense.

On the following pages are some

Guided Open-Ended Math Problems.

For each problem there are **four parts**. In the **first part**, you will solve the problem with guided help. In the **second part**, you will score and correct a solution with guided help. The **third part** shows one solution that scores a perfect **4**. This solution may or may not differ from your way. The **fourth part** has answers to the **first** and **second parts** so you can check your work.

Guided Problem #1

Brad had a $20 bill, a $10 bill, a $5 bill, and a $1 bill. Without looking, he took 2 of the bills. What is the probability that the 2 bills he took were worth at least $25?

Keywords:

1. Try It Yourself.

Answer the questions below to get a score of **4**.

What **question** are you being asked?

_____

What are the **keywords**?

_____

What are the **facts** you need to solve the problem?

_____

_____

_____

What **strategy** can you use to solve the problem?

_____

_____

Solve the problem.

_____

_____

_____

_____

Write/Explain what you did to solve the problem.

_____

_____

_____

_____

Reflect. Review and improve your work.

_____

_____

_____

Hint

Possible answers include: **Make a List** and **Make a Table**.

2. George Tries It.

George's Paper

Question: What is the probability that Brad will pick bills that equal $25 or more?

Keywords: probability, worth

Facts: Brad has 4 bills. He has one $20, $10, $5, and $1. Brad will take 2 of the bills without looking.

Strategy: Make a List.

Solve: $20 + $10 = $30
$20 + $5 = $25
$20 + $1 = $21

Write/Explain: I Made a List of the values of the bills that Brad has. There are 3 ways that Brad could pick 2 bills and 2 of them are $25 or more. The probability is $\frac{2}{3}$ that Brad will take at least $25.

Use the rubric on *page 13* to score this work.

Score the Answer.

According to the rubric from **1** to **3**, what score would you give George? Explain why you gave that score.

_____

_____

_____

_____

_____

_____

_____

Make it a 4! Rewrite.

_____

_____

_____

_____

_____

_____

_____

3. Brett Tries It.

Remember, there is often more than one way to solve a problem. Here is how Brett solved this problem.

Brett's Paper

Question: What is the probability that Brad will pick bills that equal $25 or more?

Keyword: probability

Facts: Brad has 4 bills. He has one $20, $10, $5, and $1. Brad will take 2 of the bills without looking.

Strategy: I can Make a Table.

Solve:

$20 + $10 = $30	1 for 1
$20 + $5 = $25	2 for 2
$20 + $1 = $21	2 for 3
$10 + $5 = $15	2 for 4
$10 + $1 = $11	2 for 5
$5 + $1 = $6	2 for 6

The probability of picking two bills that equal at least $25 is $\frac{1}{3}$.

Write/Explain: I had to find out how many combinations of 2 bills Brad could take. I Made a Table to show how much each combination was worth. I found there are 6 combinations and that 2 of them are equal to or greater than $25. So the chances of Brad getting more than $25 was $\frac{2}{6}$ which is the same as $\frac{1}{3}$.

Score: Brett would receive a **4** on our rubric. He showed he understood the problem perfectly. He listed the question, keywords, and facts. He chose a good strategy and applied it correctly. His answer was correct and his explanation was clear. He labeled his answer.

4. Answers to Parts 1 and 2.

Guided Problem #1

Brad had a $20 bill, a $10 bill, a $5 bill, and a $1 bill. Without looking, he took 2 of the bills. What is the probability that the 2 bills he took were worth at least $25?

Keywords: probability, worth

1. Try It Yourself. (page 99)

Question: What is the probability that Brad will pick bills that equal $25 or more?

Facts: Brad has 4 bills. He has one $20, $10, $5, and $1. Brad will take 2 of the bills without looking.

Strategy: Make a List.

Solve: $20 + $10 = $30
$20 + $5 = $25
$20 + $1 = $21
$10 + $5 = $15
$10 + $1 = $11
$5 + $1 = $6

There are 6 possible ways that Brad could take 2 bills. Two of those possibilities are for $25 or more. Therefore, the probability is $\frac{2}{6}$ or $\frac{1}{3}$ that Brad will pick 2 bills that have a value of $25 or more.

Write/Explain: I Made a List of the different ways that Brad could take 2 bills. I found the sum of each of those ways. Then I wrote a fraction to describe the probability.

2. George Tries It. (page 100)

Score the Answer: I would give George a **3** on the rubric. He knew the question that was asked, and gave the keywords and the facts. He picked a good strategy, but he did not follow through. He only listed the possibilities with picking a $20 bill. That caused him to get an incorrect answer.

Make it a 4! Rewrite.

$20 + $10 = $30
$20 + $5 = $25
$20 + $1 = $21
$10 + $5 = $15
$10 + $1 = $11
$5 + $1 = $6

There are 6 possible ways that Brad could take 2 bills. Two of those possibilities are for $25 or more. Therefore, the probability is $\frac{2}{6}$ or $\frac{1}{3}$ that Brad will pick two bills that have a value of $25 or more.

Guided Problem #2

Katie made a table showing the temperature at 12 noon for a week. What was the median temperature at 12 noon for the week?

Temperature at 12 Noon for a Week

Day	Temperature (in °C)
Sunday	14
Monday	12
Tuesday	18
Wednesday	20
Thursday	16
Friday	14
Saturday	16

Keyword:

1. Try It Yourself.

Answer the questions below to get a score of **4**.

What **question** are you being asked?

_____

What is the **keyword**?

_____

_____

What are the **facts** you need to solve the problem?

_____

_____

_____

_____

What **strategy** can you use to solve the problem?

_____

_____

Hint

Possible answers include: **Use Logical Thinking**.

Solve the problem.

_____

_____

Write/Explain what you did to solve the problem.

_____

_____

Reflect. Review and improve your work.

_____

2. Shelly Tries It.

Shelly's Paper

Question: What was the median temperature at 12 noon for the week?

Keyword: median

Facts: The temperature for each day at 12 noon is given in the table.

Solve: The middle number is the temperature on Wednesday. The median is 20.

Write/Explain: The median number is the middle number in a data set. The middle number in a set of 7 is the fourth number, which is 20.

Score the Answer.

According to the rubric, from **1** to **3**, what score would you give Shelly? Explain why you gave that score.

_____

_____

_____

_____

_____

Make it a 4! Rewrite.

_____

_____

_____

_____

_____

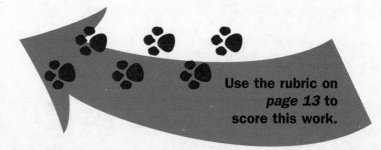

Use the rubric on *page 13* to score this work.

3. Fernando Tries It.

When solving open-ended math problems, there is often more than one way to do it. Here is how Fernando solved this problem.

Fernando's Paper

Question: What was the median temperature at 12 noon for the week?

Keyword: median

Facts: The temperature for each day at 12 noon is given in the table.

Strategy: I Made a Graph

Solve:

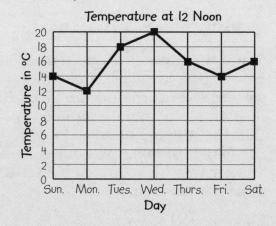

The median is the middle number in a data set. It will be the fourth number from the bottom. The median is 16°C.

Write/Explain: I Made a Line Graph. There are 7 data points, so the fourth point from the bottom is the median. The fourth number from the bottom is 16°C, so the median is 16°C.

Score: Fernando would receive a **4** on our rubric. He knew the question that was asked, and gave the keyword and the facts. Although he did a lot of extra work, his strategy led him to the correct answer. Fernando showed that he knew how to find the median of a set and labeled his answer.

4. Answers to Parts 1 and 2.

Guided Problem #2

Katie made a table showing the temperature at 12 noon for a week. What was the median temperature at 12 noon for the week?

Temperature at 12 Noon for a Week

Day	Temperature (in °C)
Sunday	14
Monday	12
Tuesday	18
Wednesday	20
Thursday	16
Friday	14
Saturday	16

Keyword: median

1. Try It Yourself. (pages 103–104)

Question: What was the median temperature at 12 noon for the week?

Facts: The temperature for each day at 12 noon is given in the table.

Strategy: Logical Thinking

Solve: Order the numbers from least to greatest:

12, 14, 14, 16, 16, 18, 20

The middle number is the median, so the median is 16°C.

Write/Explain: I used Logical Thinking. The median is the middle number in an ordered set of data. I ordered the temperatures from least to greatest. The middle number is the fourth number in this set, which was 16°C.

2. Shelly Tries It. (page 104)

Score the Answer: I would give Shelly a **2** on the rubric. She knew what question was asked, and gave the keyword and the facts. She knew that the median is the middle number, but did not know that the numbers had to be ordered. She picked the fourth number without ordering them. Her answer was incorrect. She also forgot to give her answer in degrees.

Make it a 4! Rewrite.

Order the numbers from least to greatest:

12, 14, 14, 16, 16, 18, 20

The middle number in an ordered set is the median, so the median is the fourth number, which is 16°C.

Guided Problem #3

Lynn has taken four math tests. The results are shown in the bar graph.

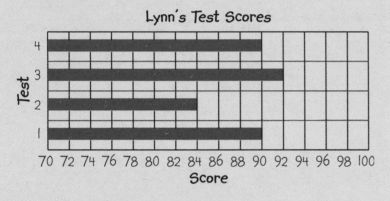

Lynn's Test Scores

What is Lynn's mean test score?

Keywords:

1. Try It Yourself.

Answer the questions below to get a score of **4**.

What **question** are you being asked?

_____

What are the **keywords**?

_____

What are the **facts** you need to solve the problem?

_____

_____

_____

_____

What **strategy** can you use to solve the problem?

_____

_____

Hint

Possible answers include **Divide and Conquer** and **Make it Simpler**.

Solve the problem.

_____

_____

_____

Write/Explain what you did to solve the problem.

_____

_____

Reflect. Review and improve your work.

_____

2. Don Tries It.

Don's Paper

Question: What is Lynn's mean score?

Keywords: bar graph, mean

Facts: Lynn's test scores are 90, 92, 84, and 90.

Solve: The number 90 is the only number to occur more than once, so it is the mean.

Write/Explain: I determined all of Lynn's test scores from the bar graph. I found the number that occurred the most, which is the mean. The mean is 90.

Score the Answer.

According to the rubric, from **1** to **3,** what score would you give Don? Explain why you gave that score.

_____

_____

_____

Make it a 4! Rewrite.

_____

_____

_____

_____

_____

Use the rubric on *page 13* to score this work.

3. A.J. Tries It.

When solving open-ended math problems, there is often more than one way to do it. Here is how A.J. solved this problem.

A.J.'s Paper

Question: What is Lynn's mean score?

Keywords: bar graph, mean

Facts: Lynn's test scores are 90, 92, 84, and 90.

Strategy: Make It Simpler.

Solve: Since all of the scores are greater than 80, I can find the difference between the scores and 80. Then find the mean of the differences and add that to 80.

$$10, 12, 4, \text{ and } 10$$
$$10 + 12 + 4 + 10 = 36$$
$$36 \div 4 = 9$$
$$80 + 9 = 89.$$

Lynn's mean score is 89.

Write/Explain: I decided to Make It Simpler. All of the scores were greater than 80, so I found how much greater than 80 each of Lynn's scores was. I then added the sum of the differences between Lynn's scores and 80, and got 36. I divided 36/4 to find the mean of the differences. Then I added the mean of 9 to 80 to get 89 for the total mean.

Score: A.J. would receive a perfect 4 on our rubric. He knew what question was being asked, and listed the keywords and the facts. He picked a good strategy and applied it correctly. His math was correct. He then explained what he did. He labeled his work. It was perfect!

4. Answers to Parts 1 and 2.

Guided Problem #3

Lynn has taken four math tests. The results are shown in the bar graph.

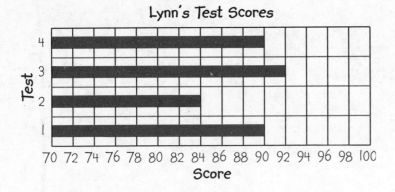

Lynn's Test Scores

What is Lynn's mean test score?

Keywords: bar graph, mean

1. Try It Yourself. (page 107)

Question: What is Lynn's mean score?

Facts: Lynn's test scores are 90, 92, 84, and 90.

Strategy: Divide and Conquer.

Solve: 90 + 92 + 84 + 90 = 356

$$
\begin{array}{r}
89 \\
4\overline{)356} \\
-32 \\
\hline
36 \\
-36 \\
\hline
0
\end{array}
$$

The mean is 89.

Write/Explain: I used **Divide and Conquer**. *First*, I found the sum of Lynn's scores. *Second*, I found the mean by dividing the sum by the number of scores.

2. Don Tries It. (page 108)

Score the Answer: I would give Don a **2**. He wrote down the correct question, the keywords, and the facts. He did not know the difference between the mean and mode. He gave an incorrect answer because he found the mode instead of the mean.

Make it a 4! Rewrite.

Find the mean by dividing the sum by the number of scores.

Find the sum: $90 + 92 + 84 + 90 = 356$.

$$
\begin{array}{r}
89 \\
4\overline{)356} \\
-\ 32 \\
\hline
36 \\
-\ 36 \\
\hline
0
\end{array}
$$

The mean is 89.

Quiz Problems

Here are some problems for you to try. Keep your **rubric** handy while you solve the problems. Let's see if you can score a **4**.

1. Daisy has scores of 78, 82, and 87 in her math tests. She has another test tomorrow. What is the least score she has to earn to have mean score of 85?

2. In a bag there are 9 marbles. Four of them are blue, 3 of them are red, and 2 of them are yellow. Without looking, Felipe picks out 1 marble from the bag. What is the probability that the marble he picked out is not blue?

3. There are 5 players entered in a table tennis tournament. For the first round, every player must play every other player 1 time. How many games will there be in the first round?

4. There are 72 campers at Robin Hood Day Camp. At 11 o'clock all of the campers picked 1 of 4 activities. The circle graph shows the results. How many campers were involved in each activity?

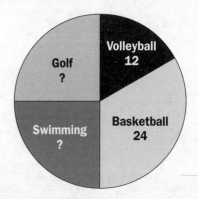

5. A potato farm recorded the weights of the potatoes sold last week. They sold 1,200 pounds of russet, 2,400 pounds of Idaho, 1,800 pounds of sweet potatoes and 1,350 pounds of Yukon gold. Make an appropriate graph to show this data. How many pounds of potatoes did they sell all together?

6. The pictograph shows the attendance at the state fair during its final week. How many more people visited the fair on Sunday than on Monday?

State Fair Attendance Final Week

Day	Number of People
Monday	△ △ △ △ △ △
Tuesday	△ △ △ △ △ △ △
Wednesday	△ △ △ △ △ △ △ △
Thursday	△ △ △ △ △ △ △ △ △ △⟋
Friday	△ △ △ △ △ △ △ △ △ △⟋
Saturday	△ △ △ △ △ △ △ △ △ △ △ △ △
Sunday	△ △ △ △ △ △ △ △ △ △ △ △ △ △ △

Each △ = 100 people

7. Jackie has been given the following data set.

35, 42, 38, 45, 35

Which of the following does 39 represent: range, mean, median, mode?

10. Test #1

1. Arlene is training for the swim meet. She plans to swim 5 laps for 3 days. Then she plans to swim 7 laps for 3 days. Then she plans to swim 9 laps for 3 days. If she continues this pattern, on which day will she first swim 15 laps?

2. Mike is playing with this function machine. What are the missing numbers?

Input	Output
1	4
2	6
3	8
4	10
5	?
6	?
8	?

3. Andie said that all rectangles are squares. Cindy said that all squares are rectangles. Who is correct? Explain your answer.

4. Luis drew a rectangle that is 10 inches long by 6 inches wide. Mitch drew a congruent rectangle. What are the dimensions of Mitch's rectangle?

5. The pictograph shows how many children volunteered to be ushers at the school show. How many more children volunteered on Friday than on Monday and Tuesday combined?

Children Who Volunteer

Day	Number of Children
Mon.	🧍🧍🧍🧍
Tues.	🧍🧍🧍
Wed.	🧍🧍🧍🧍
Thurs.	🧍🧍
Fri.	🧍🧍🧍🧍🧍🧍🧍🧍🧍

Each 🧍 = 20 children.

6. Carol pitched $\frac{7}{9}$ of a softball game. Laura pitched $\frac{2}{9}$ of the game. How much more of the game did Carol pitch than Laura?

7. Mr. Robinson gave this puzzle to his students.
"I am thinking of 2 numbers. Their product is 36 and their sum is 13. What are the 2 numbers?"

8. Bob is going to toss 3 coins at the same time. How many possible outcomes are there?

9. A rectangle has a length of 7 meters and a width of 6 meters. What is the area of the rectangle?

10. Sue, Teri, and Ursula brought home seashells. They decided to share the shells equally. Sue found 18 shells, Teri found 30 shells and Ursula found 12. How many shells did each girl take home?

11. Mrs. Johnson ordered 70 balloons, some gold and some white. She wants 20 more gold than white. How many of each will she buy?

12. A scout troop sold 3 times as many boxes of cookies Saturday as they did Friday. They sold 54 fewer boxes Sunday than they did Saturday. They sold 82 boxes of cookies Friday. How many boxes of cookies did they sell Sunday?

13. Henry arranged dimes in the shapes of 5 rectangles. The figure shows the first 3 rectangles he made. If the pattern of rectangles continues, how many dimes would be in the last rectangle?

10. Test #1

Answer the following questions to the best of your ability. Remember, even if you are unsure of how to solve the problem, you will always earn some credit if you begin the problem. Good luck!

14. Ramona spent 1 hour on math, 50 minutes on science, $\frac{1}{2}$ hour on social studies, and 25 minutes on reading. She took a 15-minute break between science and social studies. She finished at 7:00 p.m. At what time did Ramona start?

15. You can order grilled cheese, pizza, or macaroni and cheese in the school cafeteria. Students can choose either chocolate milk or apple juice. Eileen ordered pizza and apple juice. What is the probability that Sasha will order the same two items?

16. How can you find the lines of symmetry of an equilateral triangle? How many lines of symmetry does an equilateral triangle have?

17. How can the transformation of the arrows best be described?

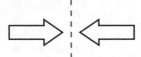

18. Mr. Lamb is tiling his den floor with vinyl tiles. The floor is a rectangle that measures 8 feet by 10 feet. Each tile is a 1-foot square. Each tile costs $1.30. How much did he spend on the tiles?

19. Pete's cat weighs 12 pounds 7 ounces. Jadine's cat weighs 160 ounces. Whose cat weighs more? How much more? Write your answer in pounds and ounces.

20. A dart board has 6 equal sections, marked with the numbers 1, 2, 3, 4, 5, and 6. Louis hit the board with a dart. What is the probability that he scored greater than 4?

10. Test #2

1. Marla is training for a cross-country race. She plans to run 3 laps of the track each day for 3 days. Then she will run 5 laps a day for 4 days. Then she will run 7 laps a day for 5 days. If she continues this pattern, what is the first day she will run 11 laps?

2. Lauren is playing with this function machine. What are the missing numbers?

Input	Output
1	1
2	3
3	5
4	7
5	?
6	?
8	?

3. What kind of triangle is always similar to every other triangle of the same type?

4. A square is made from 2 triangles. A pentagon is made from 3 triangles. A hexagon is made from 4 triangles. From how many triangles is an octagon made?

5. The pictograph shows how many trees the fourth-grade students planted.

Oak	🌳 🌳
Maple	🌳 🌳 🌳 🌳
Pine	🌳 🌳
Cedar	🌳 🌳
Fir	🌳

Each 🌳 = 10 trees

How many more oak and maple trees were planted than cedar and pine trees?

6. Nick walked for $\frac{3}{8}$ mile to school. After school he walked $\frac{1}{2}$ mile to the playground. How far did Nick walk in all?

10. Test #2

Answer the following questions to the best of your ability. Remember, even if you are unsure of how to solve the problem, you will always earn some credit if you begin the problem. Good luck!

7. Ms. Lafferty gave this puzzle to her class.

"I am thinking of 2 numbers. Their product is 48 and their difference is 8. What are the 2 numbers?"

8. Tim is going to toss a coin and a number cube with the numbers 1, 2, 3, 4, 5, and 6. What is the probability that he will toss heads and a 1?

9. A rectangle has a length of 9 centimeters and a width of 5 centimeters. What is the area of the rectangle?

10. Ian, Jerry, and Karl want to share their comic books so that each boy has the same number. Ian has 17 comics, Jerry has 13 comics, and Karl has 12 comics. How many comic books will each boy have?

11. Mrs. Jackson ordered 90 doughnuts for her party, some with icing and some without. She wants 24 more with icing than without icing. How many of each will she buy?

12. The scouts collected 4 times the weight of newspapers Saturday as they did Friday. They picked up 85 pounds less Sunday than they did Saturday. They picked up 125 pounds of newspapers Friday. How many pounds of newspaper did they pick up Sunday?

13. Alice made 6 squares out of nickels. The first square had 1 nickel. The second square had 4 nickels, the third had 9 nickels and so on. How many nickels were in the sixth square?

10. Test #2

Answer the following questions to the best of your ability. Remember, even if you are unsure of how to solve the problem, you will always earn some credit if you begin the problem. Good luck!

14. It will take Fred 45 minutes to cook the chicken once the grill is hot. It will take him 20 minutes to get the grill hot. It will take 30 minutes to prepare the chicken. He wants to serve dinner at 7:30 p.m. At what time should Fred start to prepare the chicken?

15. Danny is going to order ice cream. He can choose vanilla, chocolate, or strawberry. As a topping he can choose chocolate sprinkles, rainbow sprinkles, or whipped cream. How many possible choices does Danny have to order one flavor of ice cream with one topping?

16. How many more lines of symmetry does a square have than a rectangle? Explain your answer.

17. How can the transformation of the heart best be described?

18. Mrs. Carney is tiling her basement floor with vinyl tiles. The floor is a rectangle that measures 8 feet by 12 feet. Each tile is a 1-foot square that costs $5. How much did she spend on the tiles?

19. Tony is 5 feet 4 inches tall. He is 17 inches taller than Pam. How tall is Pam? Write your answer in feet and inches.

20. Tari had 4 balls in a bag. The balls were numbered 1, 2, 3, and 4. She reached into the bag and pulled out 2 of the balls. Then she found their sum. What is the probability that the sum was even?

11. Home-School Connection

Working on these questions at home with a family member is fun! Find a comfortable place to work and have all the tools you need. Discuss how you want to solve the open-ended math question. Then go for it! Don't forget to use your rubric!

Dear Family Member:

This year your child will be learning about **open-ended questions** in mathematics class. An open-ended math question is a mathematics word problem that has one correct answer, but that can be solved in several different ways. Open-ended math questions are extremely important on tests your child will take.

You can help your child practice solving these questions by working together on the take-home sheets in this chapter. Don't forget to use the rubric as a guide. Remember, when you work with your child, don't do the problem. Rather, encourage your child to ask questions that will lead him or her to the answer. And don't be surprised if your child arrives at the answer to the problem using a method different from the one you're thinking of.

It is important that your child make his or her thinking clear to the reader. After your child has solved the problem, (or when you child has gone as far as he or she can) help your child write a clear explanation of what he or she did to solve it, and why he or she decided to do it that way. This will help your child clarify his or her own thoughts.

The problems on the following pages are based on the areas of mathematics considered important in solving open-ended math problems. These are:

- **Number and Operations**

- **Algebra**

- **Geometry**

- **Measurement**

- **Data Analysis and Probability**

Enjoy!

Number and Operations

Problem

Zach went to the supermarket for his mother. He bought 5 pounds of potatoes at 39¢ a pound and 2 loaves of bread at $1.69 each. How much change will he receive from a $10 bill?

Algebra

Problem

Jonathan and Max went to the bookstore. Together they spent $21 on books.

Jonathan spent $5 more than Max. How much did each boy spend?

Geometry

Problem

Ms. Park described this mystery figure to the students in her class.

"The figure I am thinking of has 3 dimensions and 5 faces. The figure is made of 4 triangles and a rectangle. What figure am I thinking of?" Explain your answer.

Measurement

Problem

For a fruit punch, Mrs. Abramson used 3 quarts of orange juice, 4 pints of apple juice, and $\frac{1}{2}$ gallon of pineapple juice. How many cups of juice were used to make the fruit punch?

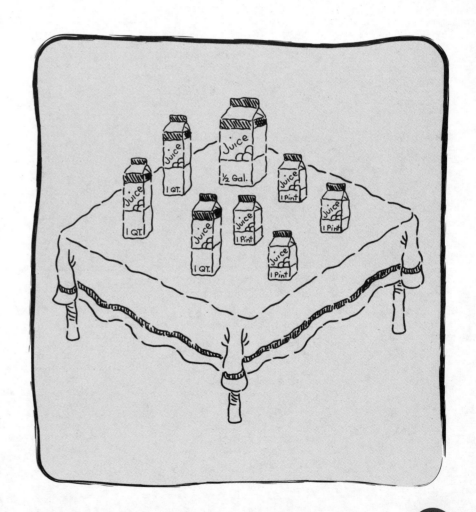

Data Analysis and Graphs

Problem

Today the lunch choices for main courses are hamburger, grilled cheese, or pizza. The choices for dessert are an apple, an orange, or a slice of watermelon. How many different lunches of one main course and one dessert can a student buy?

Glossary

A

Add   To combine two or more numbers to find a sum.

Addend   One of two or more numbers that are added.

Algebra   A branch of mathematics that uses letters and symbols to write expressions about number relationships.

Alike   Same.

All together   With all included.

Always   Every time.

Angle   The figure formed by two lines that start at one point.

Appropriate   Fitting, suitable.

Area   The number of square units needed to cover a region.

Glossary

B

Base A side of a polygon, usually the one at the bottom.

Best Most appropriate.

Boundary A line or any other thing that fixes a limit.

C

Center The middle.

Change The amount of money that is given back after a purchase.

Chord A line segment that connects two points on a circle.

Circle graph A graph in which data are represented by parts of a circle. It shows parts of a whole.

Classification An arrangement in groups based on established criteria.

Congruent

Having the same size and shape as another figure.

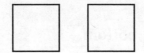

These squares are congruent.

Coordinate grid

A grid used to show location.

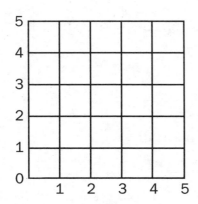

D

Degree

Customary unit for measuring angles.

Degree

Unit for measuring temperature.

  Celsius (°C)

Metric unit for measuring temperature; freezing point: 0°C; boiling point: 100°C

  Fahrenheit (°F)

Customary unit for measuring temperature; freezing point: 32°F; boiling point: 212°F

Diagram

A drawing that explains something.

Diameter   A line segment that passes through the center of a circle.

Difference   The answer in a subtraction problem.

Different   Not the same.

Digits   Numerals from 1 to 9, and sometimes 0.

Dimension   One of a group of properties such as length, width, and height.

E

Each   Every one of a group considered individually.

Enough   An adequate quantity.

Equally share   To divide into equal groups.

Equilateral triangle   A triangle with 3 congruent sides and angles.

Even number   A number that has 0, 2, 4, 6, or 8 in the ones column.

Every   All.

Expire   End.

Expression   A mathematical statement made from mathematical symbols. Ex. $6 + 3$, $7 > 3$, $9 = 9$.

F

Fallen   Decreased.

Fewer   Less than another number.

Figure   A shape, or an amount, given in numbers.

Finished   Ended.

First   Preceding all others.

Fluid ounce   A customary unit of capacity. Eight fluid ounces is equal to a cup.

Foot   A customary unit of length equal to 12 inches. Plural is feet.

Function   A relationship in which one quantity depends on another quantity.

Glossary

G

Geometry The study of figures.

Graph A picture that displays the data.

Guess and Test A problem-solving strategy in which you make a guess, test your guess, and then change your guess until finding the correct answer.

H

Half One of two equal parts.

Height The measurement of how high (the greatest point) something is.

Hexagon A polygon with 6 sides and 6 angles.

High Measuring from top to bottom. The height of something.

Hour A unit of time that contains 60 minutes.

I

In all   With all included.

Inch   A customary unit of length equal to $\frac{1}{12}$ of a foot.

Isosceles triangle   A triangle with 2 sides that are the same length and 2 equal angles.

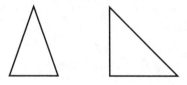

L

Least   The smallest number in a set.

Length   The measurement of distance between two endpoints.

Less   Fewer.

Line   A long, thin continuous mark.

Line of symmetry   A line on which a figure can be folded so that both sides match.

Long   Length from one end to the other.

Glossary

M

Many   A large number.

Mean   A measure of the sum of the addends divided by the number of addends.

Measure   To find out the size, length, weight, etc. of something.

Median   The middle number in an ordered set.

Minute   A unit of time equal to 60 seconds.

Missing   Not present.

Mode   The number that occurs the most in a data set.

Month   A period of time lasting about 30 days. There are 12 months of the year.

More   Greater in number or amount.

Most   More than $\frac{1}{2}$.

Much   Large in number.

N

Net A 2-dimensional pattern that can be folded to make a 3-dimensional figure.

Not A function word used to make negative a group of words or a word.

Number A word or symbol used for counting, amounts, and other mathematical needs.

Number sentence An equation.

O

Once One time.

Open-ended problem A math problem that has a correct answer that you can arrive at in more than one way.

Opposite Located or facing directly across, or completely different.

Ordered pair A pair of numbers that gives the location on a coordinate grid.

Ounce A customary unit of weight equal to $\frac{1}{16}$ of a pound.

Glossary

P

Parallelogram   A polygon with 4 sides with opposite sides parallel.

Pattern   A series of numbers or figures that follows a rule.

Pentagon   A polygon with 5 sides and 5 angles.

Perimeter   Distance around the edge of a shape or an area.

Pictograph   A graph that compares data by using pictures or symbols.

Pint   A customary unit of capacity equal to 2 cups.

Point   An exact location in space.

Polygon   A figure with three or more straight sides.

Possible outcomes   Each of the outcomes of a probability experiment.

148

Pound	A customary unit of weight equal to 16 ounces.
Probability	The chance of an event occurring.
Product	The answer in a multiplication problem.

Q

Quadrilateral	A figure with four sides.
Quart	A customary unit of capacity equal to 2 pints.

R

Range	The difference between the greatest number and the least number in a data set.
Rectangle	A polygon with four sides, four right angles, and two pairs of opposite congruent sides.

Rectangular	Having the shape of a rectangle.

Glossary

Reflection Flip. A movement of a figure to a new position by flipping the figure over a line.

Rotation A movement of a figure to a new position by rotating the figure around a point.

Rotational symmetry A figure has rotational symmetry if it can be turned and match itself with a $\frac{1}{2}$ turn or less.

S

Same Equal to.

Scalene triangle A triangle whose 3 sides each have different lengths.

Segment A part or section of something.

Shape A form or outline of an object or figure.

Side One of the line segments that make up a polygon.

Similar   Figures that have the same shape, but possibly different sizes.

These squares are similar.

Specifically   As exactly as possible.

Square   A rectangle with four equal sides.

Start   The beginning.

Sum   The answer in an addition problem.

T

Temperature   A measurement that tells how hot or cold something is.

Three-digit number A whole number with a digit in the hundreds, tens, and ones places.

Glossary

Three-dimensional figure   A figure that has length, width, and height.

Times   Multiplied by.

Together   Combined.

Total   Sum.

Transformation   The movement of a figure. Transformations include translations, rotations, and reflections.

The movement of a figure by a 1- slide, 2- flip, or 3- turn.

Examples:

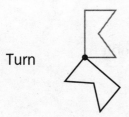

Slide

Turn

Flip

152

Translation   A movement of a figure to a new position without turning or flipping it.

Triangle   A polygon with three sides and three angles.

Twice   Two times.

U

Unit   A single thing or amount.

Until   Continuing up to the end of an activity. Up to the point of.

V

Vertex   The point at which the sides of an angle intersect or meet.

Volume   The amount of space a three-dimensional figure holds.

W

Wide
The distance from one side of something to the other. Width.

Width
The measurement of how wide something is.
The length of the shorter dimension of a polygon.

Worth
Having a value in money.

Math Abbreviations

centimeter	cm	hour	h	milligram	mg	pound	lb
cup	c	inch	in.	milliliter	mL	quart	qt
day	d	kilogram	kg	millimeter	mm	second	s
fluid ounce	fl oz	kilometer	km	minute	min	ton	T
foot	ft	liter	L	month	mo	week	wk
gallon	gal	meter	m	ounce	oz	yard	yd
gram	g	mile	mi	pint	pt	year	y

Days of the Week	Months of the Year	Days in Month
Sunday	January	31
Monday	February	28 or 29
Tuesday	March	31
Wednesday	April	30
Thursday	May	31
Friday	June	30
Saturday	July	31
	August	31
	September	30
	October	31
	November	30
	December	31

Larger Units of Time

1 week (wk): 7 days

1 month (mo): about 30 days

1 year (y): 12 months or 365 days

1 year: about 52 weeks

1 leap year: 366 days

1 decade: 10 years

1 century: 100 years

Math Symbols

+	addition	"	inches
−	subtraction	'	feet
×	multiplication	°C	degrees Celsius
÷	division	°F	degrees Fahrenheit
=	is equal to	$\overleftrightarrow{AB}$	line AB
>	is greater than	$\overrightarrow{AB}$	ray AB
<	is less than	$\overline{AB}$	line segment AB
·	decimal point	∠A	angle A
$	dollars	△ABC	triangle ABC
¢	cents	(2, 3)	ordered pair (2, 3)

Notes

Notes